Reactions at solid surfaces

REACTIONS AT SOLID SURFACES

GERHARD ERTL
Fritz-Haber-Institut der
Max-Planck-Gesellschaft
Berlin, Germany

A JOHN WILEY & SONS, INC., PUBLICATION

Copyright © 2009 by John Wiley & Sons, Inc. All rights reserved.

Published by John Wiley & Sons, Inc., Hoboken, New Jersey.
Published simultaneously in Canada.

No part of this publication may be reproduced, stored in a retrieval system, or transmitted in any form or by any means, electronic, mechanical, photocopying, recording, scanning, or otherwise, except as permitted under Section 107 or 108 of the 1976 United States Copyright Act, without either the prior written permission of the Publisher, or authorization through payment of the appropriate per-copy fee to the Copyright Clearance Center, Inc., 222 Rosewood Drive, Danvers, MA 01923, (978)-750-8400, fax (978)-750-4470, or on the web at www.copyright.com. Requests to
 the Publisher for permission should be addressed to the Permissions Department, John Wiley & Sons, Inc., 111 River Street, Hoboken, NJ 07030, (201)-748-6011, fax (201)-748-6008, or online at http://www.wiley.com/go/permissions.

Limit of Liability/Disclaimer of Warranty: While the publisher and author have used their best efforts in preparing this book, they make no representations or warranties with respect to the accuracy or completeness of the contents of this book and specifically disclaim any implied warranties of merchantability or fitness for a particular purpose. No warranty may be created or extended by sales representatives or written sales materials. The advice and strategies contained herein may not be suitable for your situation. You should consult with a professional where appropriate. Neither the publisher nor author shall be liable for any loss of profit or any other commercial damages, including but not limited to special, incidental, consequential, or other damages.

For general information on our other products and services, or technical support, please contact our Customer Care Department within the United States at (800)-762-2974, outside the United States at (317)-572-3993 or fax (317)-572-4002.

Wiley also publishes its books in a variety of electronic formats. Some content that appears in print may not be available in electronic books. For more information about Wiley products, visit our Web site at http://www.wiley.com.

Library of Congress Cataloging-in-Publication Data:

Ertl, G. (Gerhard)
  Reactions at solid surfaces / Gerhard Ertl.
      p. cm.
  Includes bibliographical references and index.
  ISBN 978-0-470-26101-9 (cloth)
1. Surface chemistry. I. Title.
  QD506.E775 2009
  541'.33--dc22
                                    2009028884

Printed in the United States of America

10 9 8 7 6 5 4 3 2 1

Contents

Preface, ix

1. Basic principles, 1
    1.1. Introduction: The surface science approach, 1
    1.2. Energetics of chemisorption, 4
    1.3. Kinetics of chemisorption, 11
    1.4. Surface diffusion, 13
    References, 17

2. Surface structure and reactivity, 21
    2.1. Influence of the surface structure on reactivity, 21
    2.2. Growth of two-dimensional phases, 24
    2.3. Electrochemical modification of surface structure, 29
    2.4. Surface reconstruction and transformation, 33
    2.5. Subsurface species and compound formation, 42
    2.6. Epitaxy, 44
    References, 47

3. Dynamics of molecule/surface interactions, 51
    3.1. Introduction, 51
    3.2. Scattering at surfaces, 52
    3.3. Dissociative adsorption, 54
    3.4. Collision-induced surface reactions, 59

- 3.5. "Hot" adparticles, 60
- 3.6. Particles coming off the surface, 64
- 3.7. Energy exchange between adsorbate and surface, 69
- References, 75

4. **Electronic excitations and surface chemistry, 79**
   - 4.1. Introduction, 79
   - 4.2. Exoelectron emission, 81
   - 4.3. Internal electron excitation: "chemicurrents", 86
   - 4.4. Electron-stimulated desorption, 88
   - 4.5. Surface photochemistry, 94
   - References, 98

5. **Principles of heterogeneous catalysis, 103**
   - 5.1. Introduction, 103
   - 5.2. Active sites, 105
   - 5.3. Langmuir–Hinshelwood versus Eley–Rideal mechanism, 109
   - 5.4. Coadsorption, 111
   - 5.5. Kinetics of catalytic reactions, 113
   - 5.6. Selectivity, 117
   - References, 120

6. **Mechanisms of heterogeneous catalysis, 123**
   - 6.1. Synthesis of ammonia on iron, 123
   - 6.2. Synthesis of ammonia on ruthenium, 134
   - 6.3. Oxidation of carbon monoxide, 139
   - 6.4. Oxidation of hydrogen on platinum, 149
   - References, 154

7. **Oscillatory kinetics and nonlinear dynamics, 159**
   - 7.1. Introduction, 159
   - 7.2. Oscillatory kinetics in the catalytic CO oxidation on Pt(110), 163
   - 7.3. Forced oscillations in CO oxidation on Pt(110), 169
   - References, 172

8. SPATIOTEMPORAL SELF-ORGANIZATION IN SURFACE REACTIONS, 175

    8.1. INTRODUCTION, 175
    8.2. TURING PATTERNS AND ELECTROCHEMICAL SYSTEMS, 178
    8.3. ISOTHERMAL WAVE PATTERNS, 183
    8.4. MODIFICATION AND CONTROL OF SPATIOTEMPORAL PATTERNS, 189
    8.5. THERMOKINETIC EFFECTS, 195
    8.6. PATTERN FORMATION ON MICROSCOPIC SCALE, 198
    REFERENCES, 200

INDEX, 205

PREFACE

Professors H. Abruña and M. Hines kindly invited me to deliver the 2007 Baker Lectures at the Department of Chemistry and Chemical Biology of Cornell University. Hence, in the spring that year, my wife and I spent a few weeks in Ithaca, New York, where I presented a series of lectures to people of different scientific backgrounds. We are very grateful to our hosts and all members of the department who made this stay so pleasant and inspiring. When I was asked afterward to write a book based upon the lectures for John Wiley & Sons, it was a pleasure for me to accept this request. The text herewith closely follows the eight lectures that were delivered at the 2007 Baker Lecture Series, and the content presented is essentially based on results obtained in the author's own laboratory. That is why it is not a comprehensive review, but rather a subjective picture of the field covered, reactions at solid surfaces. I have to, therefore, apologize for the fact that important work by other researchers will be inadequately represented.

I am very much indebted to my numerous coworkers who collaborated with me over many years. In addition, I am very grateful to Waruno Mahdi for careful preparation of the figures and to Marion Reimers for typing the text.

GERHARD ERTL
Berlin, November 2008

CHAPTER 1

BASIC PRINCIPLES

1.1. INTRODUCTION: THE SURFACE SCIENCE APPROACH

A solid body is always terminated by surfaces where the atoms have a different environment (e.g., fewer nearest neighbors) from that in the bulk. As a consequence, these surface atoms will exhibit altered chemical reactivity. Unsaturated valencies will give rise to bond formation with particles impinging from the adjacent (gaseous or liquid) phase, and these "chemisorbed" species will in turn differ in reactivity from that in the absence of the surface. This is the basic principle underlying the phenomenon of heterogeneous catalysis. Deposition of material beyond the first monolayer leads to nucleation of a new phase and eventually to crystal growth (epitaxy). Control of these processes on the nanometer scale is of crucial importance, for example, for semiconductor microtechnology, and the whole field of "nanotechnology" is in fact essentially governed by surface reactions. Atoms can, on the other hand, also be removed from the surface, either thermally or, if this process is associated with charge transfer across the interface,

---

Reactions at Solid Surfaces. By Gerhard Ertl
Copyright © 2009 John Wiley & Sons, Inc.

with the aid of a proper electric potential. These electrochemical reactions are underlying the processes of etching or corrosion.

This text is intended to outline our present understanding of the fundamental processes underlying reactions at solid surfaces instead of attempting to provide a full overview. For this reason, the discussion will essentially be restricted to the simplest situations: processes occurring only in two dimensions, that is, involving chemisorbed phases, on surfaces consisting of only one element, that is, metals. This scenario is found with a large variety of heterogeneously catalyzed reactions for which a few case studies will be discussed later.

Since the rate of such a reaction is proportional to the area of the exposed surface, catalysts generally exhibit a high specific surface area. Apart from the use of highly porous materials with large "internal" surface areas (e.g., zeolites), this is mostly achieved by depositing small particles of the active catalyst material onto (more or less) inert high surface area supports. Figure 1.1 shows a high-resolution electron micrograph of a Ru catalyst on a MgO support together with a cartoon illustrating the different crystal planes and edge atoms acting as active sites [1]. The catalyst particles have indeed diameters of only a few nanometers or even less: In fact, heterogeneous catalysis has been a nanotechnology for more than a hundred years, long before this term was introduced. Metal particles consisting only of a very small number of atoms may exhibit electronic properties and hence chemical reactivity different from those of the bulk material. A prominent example for this effect is offered by gold: While the bulk material is catalytically practically inert, very small particles or thin films may exhibit extraordinary activities [2], and this is a field of great current interest. However, alterations of the bulk electronic properties of the catalyst particles will be ignored in the following.

But there is another effect that may have utmost influence on the reactivity: Small catalyst particles exhibit different crystal planes together with structural defects and chemisorbed foreign

Introduction: The Surface Science Approach

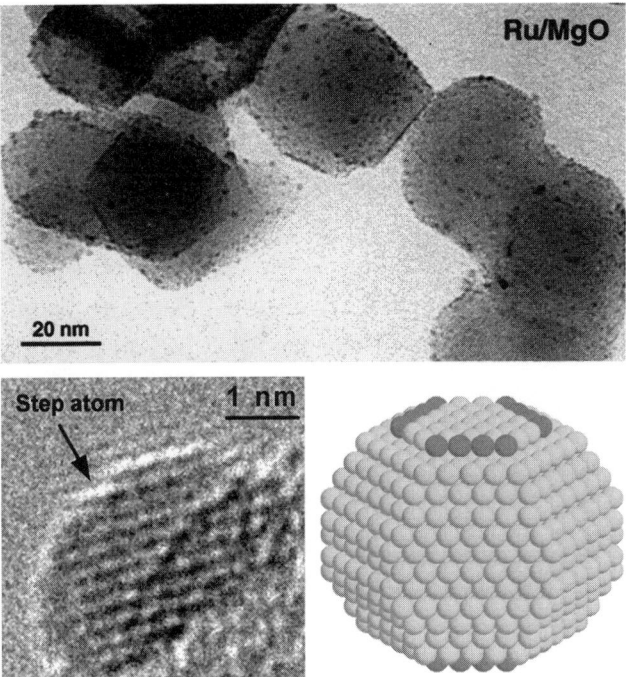

FIGURE 1.1. High-resolution electron micrograph from a small Ru particle on a MgO support together with a sketch of its structure [1].

atoms. All these effects render the surface chemistry of a "real" catalyst rather complex. A solution to this problem was already proposed by Langmuir [3] in 1922:

> Most finely divided catalysts must have structures of great complexity. In order to simplify our theoretical consideration of reactions at surfaces, let us confine our attention to plane surfaces. If the principles in this case are well understood, it should then be possible to extend the theory to the case of porous bodies. In general, we should look upon the surface as consisting of a checkerboard ...

What Langmuir had in mind were clean, well-defined single-crystal surfaces that can now be prepared and investigated through the introduction of ultrahigh vacuum techniques and the development of a whole arsenal of surface physical methods.

Since the latter in most cases cannot be operated at the high-pressure conditions of "real" catalysis, this causes the appearance of a "pressure gap." And since the properties of well-defined single-crystal surfaces will generally be quite different from the surface properties of "real" catalysts, this gives rise to the so-called "materials gap." That these gaps can indeed be overcome will be demonstrated by some of the examples to be presented.

One of the leading researchers of "classical" catalysis expressed his opinion about this "surface science approach" as follows [4]: "Catalysis is a kinetic phenomenon. The urgent need for rate constants demands the support of surface science."

The physical tools for chemical analysis of surfaces as well as for investigation of their structural, electronic, vibrational, and dynamic properties have been described quite extensively in the literature [5–11], so we refrain here from repetitions. Scanning probe techniques, and in particular the scanning tunneling microscope [12], proved to be most powerful for direct observation of processes on atomic scale.

1.2. ENERGETICS OF CHEMISORPTION

Apart from ubiquitous van der Waals interactions leading to a weak physisorption bond, particles impinging onto a solid surface may experience chemical bond formation called chemisorption—a concept originally suggested by Haber [13] and somewhat later substantiated by Langmuir [14]. This bond formation may keep the molecular entity intact (nondissociative chemisorption), or it may be associated with bond breaking and separation of the fragments on the surface (dissociative chemisorption). The reverse processes are called desorption. The strength of the chemisorption bond (i.e., chemisorption energy) may be directly determined by calorimetry. Recent developments even provide such data from single crystals, but these techniques are elaborate and hence applied only in a few laboratories [15,16]. If adsorption is in equilibrium with desorption, determination of the coverage Θ as a function of partial pressure p and temperature T provides E_{ad} through application of the Clausius–Clapeyron equation

$$\left.\frac{d\ln p}{d(1/T)}\right|_{\Theta=\text{const}} = -\frac{E_{ad}}{R}$$

This means a plot of $\ln p$ over $1/T$ at constant coverage Θ yields the isosteric heat of adsorption at the respective coverage. As an example, Fig. 1.2 shows the variation of E_{ad} for CO adsorbed on Pd(1 1 1) with Θ as determined in this way, where the coverage was monitored through the respective change in the work function [17]. The adsorption energy remains constant up to $\Theta = 0.33$ and then drops by 2 kcal/mol due to a change in the adsorption geometry as a consequence of the onset of repulsions between the adsorbed molecules. The full line in Fig. 1.2 shows the variation of the adsorption energy with coverage (i.e., mean distance between the adsorbed molecules) if the (slightly modified) interaction potential between free CO molecules is operating, which fits perfectly the experimental data at high coverages.

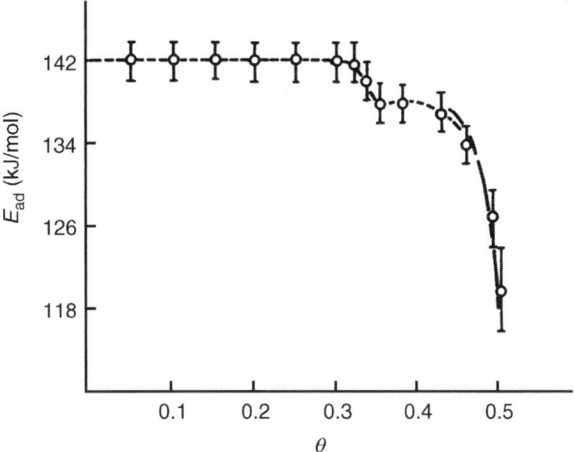

FIGURE 1.2. The adsorption energy for CO adsorbed on a Pd(1 1 1) surface as a function of coverage θ [17].

In general, interactions between adsorbates may be either repulsive or attractive. Direct repulsive interactions result from dipole–dipole interaction or from orbital overlap, but may, however, also be of indirect nature mediated through the electronic system of the substrate [29]. Attractive interactions are usually of the latter type and are analogous to the through-bond interactions in organic chemistry [18]. Figure 1.3 shows the variation of the O–O interaction potential with distance on Ru(0001) as determined through the mean residence times of the adsorbed O atoms in different configurations [19].

The most convenient (but also least accurate) method to derive information about the adsorption energy is based on the analysis of thermal desorption spectroscopy (TDS) data [5,20]. The temperature of the adsorbate covered surface is increased continuously with a constant heating rate β (so the momentary surface

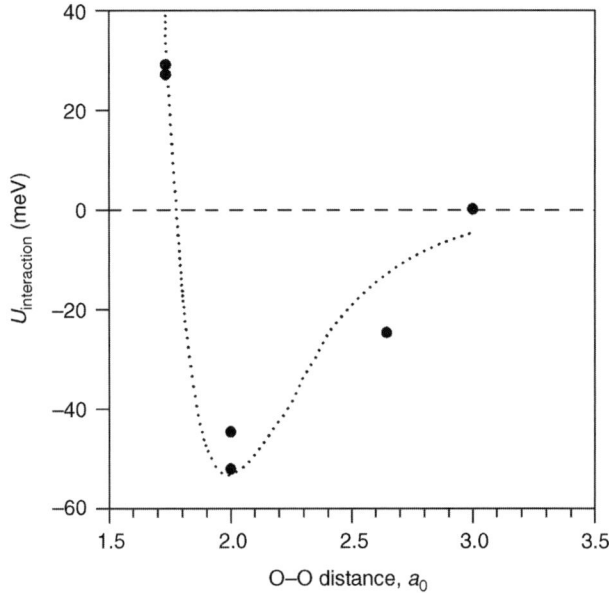

FIGURE 1.3. Variation of the interaction potential between two O atoms adsorbed on Ru(0001) as a function of their separation (a_0 = lattice constant of the substrate surface) [19].

temperature is $T = T_0 + \beta t$), and the concentration of desorbing species is monitored by a quadrupole mass spectrometer, which at high pumping rate is proportional to the rate of desorption:

$$-\frac{dn_i}{dt} = \nu_i n_i^x \exp\left(-\frac{E_{ides}^*}{RT}\right)$$

Here ν_i is the frequency factor ("preexponential"), x is the reaction order, and E_i^* is the activation energy for desorption. If adsorption is nonactivated, the latter quantity equals E_{ad}. The main problem lies in the fact that ν_i and x are usually unknown, so a simple determination of E_{ad} from the TDS peak temperature T_{max} [21] has to rely on reasonable assumptions of these quantities. More reliable determination has to be based on analysis of TDS peak shapes [5,22]. The preexponential ν may by regarded as representing the frequency of vibration of the adsorbed particle against the surface and is frequently assumed to be of the order of 10^{-13} s, but may actually deviate from this value by up to several orders of magnitude.

The energetics of dissociative adsorption can readily be rationalized by means of the one-dimensional potential diagram proposed by Lennard-Jones [23] and reproduced in Fig. 1.4: If a diatomic molecule A_2 approaches a surface, it will first experience (weak) bonding as $A_{2,ad}$. Dissociation of the free molecule would require the dissociation energy E_{diss}, and the two atoms would then form strong bonds with the surface (A_{ad}). The crossing point of the two lines marks the activation energy for dissociative adsorption and determines the kinetics of adsorption (see below), while the adsorption energy E_{ad} for $A_2 \rightarrow 2A_{ad}$ is related to the surface–adsorbate bond energy E_{S-A} through $E_{S-A} = \frac{1}{2}(E_{ad} + E_{diss})$. In the case of noninteracting adsorbed species A_{ad}, desorption then follows second-order kinetics and the TDS traces are characterized by a shift of the peak maxima to lower temperatures with increasing coverage as can be seen from Fig. 1.5 with data from the H_2/Ni (1 0 0) system [24].

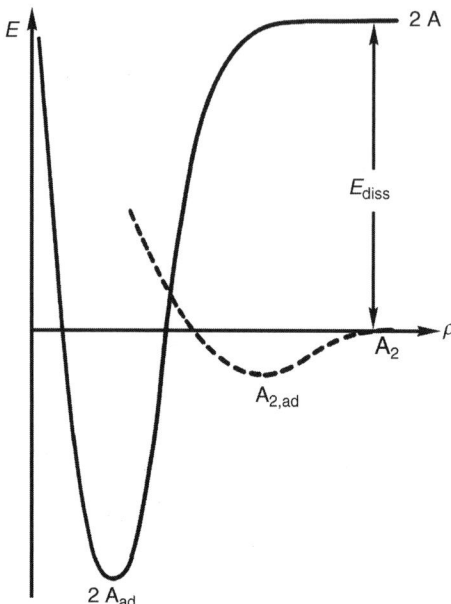

FIGURE 1.4. Lennard-Jones diagram illustrating the energetics of dissociative adsorption.

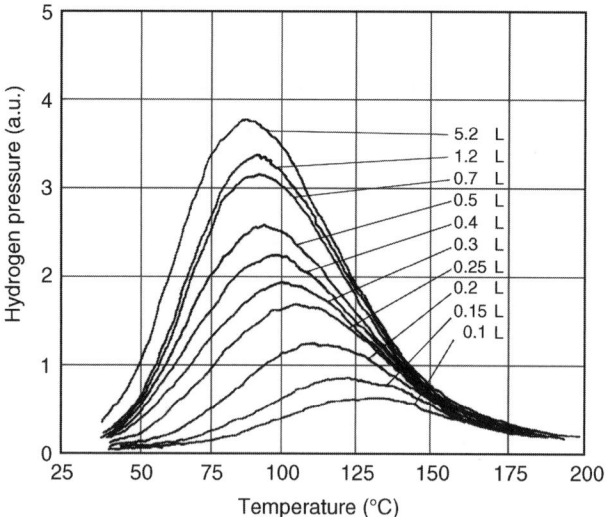

FIGURE 1.5. A series of (second-order) thermal desorption spectra for recombinative desorption of H_2 from an Ni(100) surface. Parameter is the initial exposure in Langmuir [24].

TABLE 1.1. M–CO BOND ENERGIES (KJ/MOL)

W(110)	113	Ru(0001)	121	Ir(111)	142	Ni(111)	113
W(CO)$_6$	180	Ru$_3$(CO)$_{12}$	171	Ir$_4$(CO)$_{12}$	188	Ni(CO)$_4$	146

Chemisorption is essentially a localized phenomenon, involving mainly the adsorbate and the neighboring atoms of the adsorption site. Table 1.1 lists some data for the energy of CO chemisorption on the most densely packed planes of some transition metals together with the M–CO dissociation energies of corresponding carbonyl compounds [25]. The values for the compounds are typically larger by about 30–50%, regardless of whether mono- or multinuclear carbonyls are considered. This finding appears to be qualitatively plausible, since a surface atom is always surrounded by a larger number of neighboring atoms and therefore exhibits reduced free valency. This is confirmed by the fact that the chemisorption energy is usually higher on crystallographically more open planes. Also defects, such as monoatomic steps, are associated with higher adsorption energies, and the effect of surface structure is typically of the order of about ±10%. By moving an adsorbate across the surface, the chemisorption energy varies by a similar order of magnitude. This difference determines the activation energy for surface diffusion, which is therefore typically smaller than about 20% of the adsorption energy, so the adsorbed particle makes many jumps across the surface before it eventually desorbs.

Relations between coordination chemistry of single metal atoms and surface chemistry are illustrated by the interaction of H$_2$ with either a Ru atom or a RuO$_2$(110) single-crystal surface [26]. As shown in Fig. 1.6, H$_2$ forms weakly held η^2-H$_2$ complexes with transition metal atoms [27], while on RuO$_2$(110) the H$_2$ molecule is held in a similar way above the Ru atoms, where bond lengths, vibrational frequencies, and bond strengths are quite similar in both cases. However, the further reactivity is different: While with the complex compound, dissociation of the

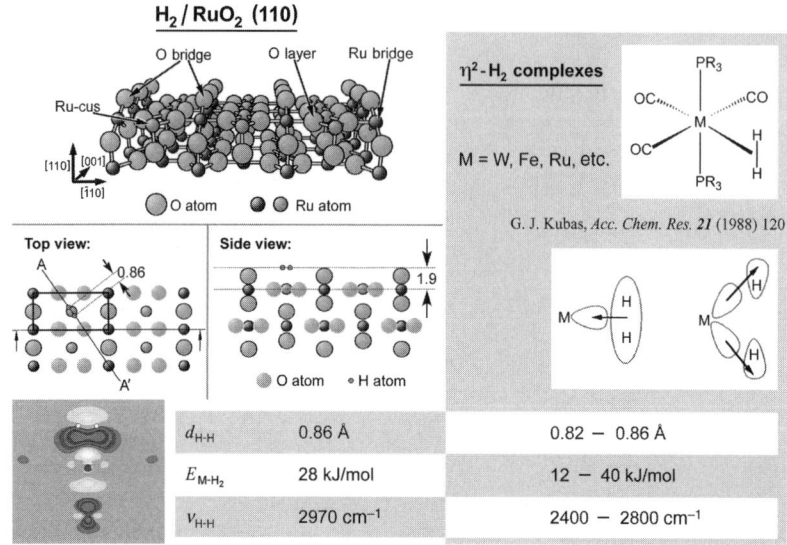

FIGURE 1.6. The bonding of H_2 on a single Ru atom or on a $RuO_2(110)$ surface [26]. (See color insert.)

H_2 molecule leaves both H atoms attached to the central metal atom, with the surface the H atoms prefer to become attached to a neighboring atom into a dihydride configuration.

As far as data are available, with transition metals the metal–metal bond energies are quite similar in cluster compounds and in bulk metals, but—what is even more important—are also comparable to the strength of the chemisorption bond with, for example, CO. In this way, it can be rationalized why the structure of a metal surface is frequently affected by chemisorption.

Theoretical description of the chemisorption bond and calculation of adsorption energies are nowadays mainly based on application of density functional theory (DFT) [28]. This approach has developed to a computational strategy of comparable accuracy to the traditional correlated quantum chemical methods, but at much lower costs, and is now widely used to calculate bond energies to fairly high accuracy comparable to experimental data [29], but sometimes also at variance [30].

1.3. KINETICS OF CHEMISORPTION

Upon adsorption the coverage of the surface by the adsorbate changes, where the *absolute* coverage Θ is defined as the ratio of the density of adsorbed particles n_a to the density of surface atoms in the topmost layer n_s, $\Theta = n_a/n_s$. Saturation equals only in rare cases $\Theta = 1$, so this definition is at variance with the original Langmuir picture [31] assuming that each surface atom represents an adsorption site. This fact is taken into account by introducing the *relative* coverage $\delta = \Theta/\Theta_{sat}$, which then reaches 1 at saturation.

The flux of particles impinging on the surface per cm^2 per second is given by

$$f_s = \frac{p}{\sqrt{2\pi m k_B T}}$$

where p is the pressure (Pa), m is the mass of the incident particle (kg), k_B is Boltzmann's constant, and T is the absolute temperature. As a rule of thumb, 1×10^{-6} Torr impinging on the surface for 1 s would suffice to completely cover the surface if each particle striking the surface is adsorbed. The exposure of 10^{-6} Torr s is denoted as 1 L (Langmuir).

Only in rare cases each particle striking the surface will become adsorbed, but only a fraction s, called the sticking coefficient. Generally, s will decrease with increasing coverage from its initial value s_0 in the simplest (Langmuir) case of nondissociative adsorption as $s = s_0(1 - \delta)$. This is the simplest case that assumes that whenever a particle strikes an empty site it will be adsorbed with probability s_0, otherwise it is reflected and the adsorbates are randomly distributed on the surface.

An extension of this approximation is the so-called "precursor" model [32,33], which is illustrated in Fig. 1.7. This model assumes a finite lifetime of particles in a second layer on the top of the chemisorbed phase ("extrinsic precursor") during

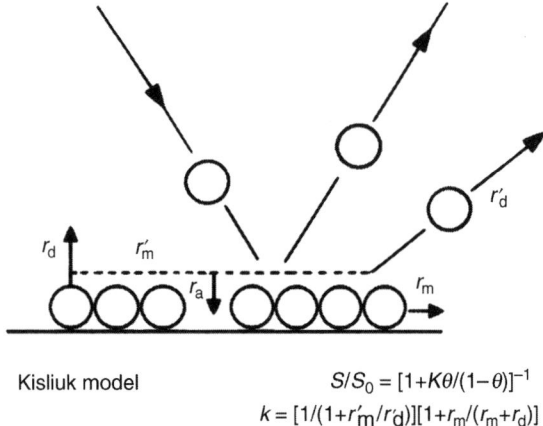

FIGURE 1.7. The Kisliuk model for "precursor" mediated adsorption.

which they may reach unoccupied chemisorption sites or otherwise desorb. As a consequence, at lower coverages the decrease of s with δ becomes flatter than that with the Langmuir model. An example for such a behavior is shown in Fig. 1.8 for the system CO/Pd(1 1 1) [34]. In this case, the initial sticking coefficient $s_0 = 0.96$ is close to unity. These data were obtained by recording

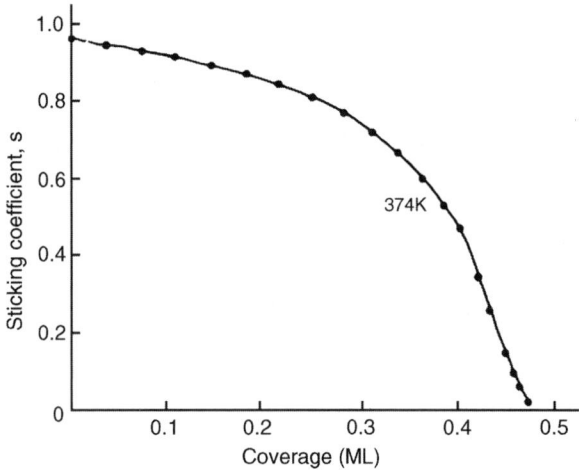

FIGURE 1.8. The sticking coefficient for CO on Pd(1 1 1) as a function of coverage θ [34].

$D_x^* = \langle x^2 \rangle/2t$, where $\langle x^2 \rangle$ is the mean square displacement of a randomly walking particle after time t. Experimental verification of this equation is, however, rather difficult. The reason is that determination of the Fickian diffusion coefficient by following the decay of the concentration profile of an adsorbed species requires naturally finite coverages where interactions between adsorbed particles come into play, so the general relation between the Fickian, or chemical, diffusion constant D_x and D_x^*, derived from the random motion of individual particles, is given by $D_x = D_x^*[\partial(\mu/kT)/\partial \ln \Theta]_T$ (where μ is the chemical potential) [37], where the factor in brackets becomes unity only for $\Theta \to 0$. Another problem with macroscopic measurements is that even well-prepared single-crystal surfaces contain numerous defects with different adsorbate binding properties. Individual particle hopping can be followed apart from field ion microscopy (FIM) [38] by scanning tunneling microscopy (STM) [39,40].

The latter technique was also applied for demonstrating the equality of D_x and D_x^* (in the absence of adsorbate–adsorbate interactions) with N atoms diffusing on a Ru(0 0 0 1) surface [40]. Exclusive dissociation of NO at 300K at monoatomic steps creates a quasi δ-function of N_{ad} concentration at $t = 0$. Determination of their mean square displacement away from the step as a function of time leads to a straight line whose slope yields $D_x^* = (3.4 \pm 0.4) \times 10^{-18}$ cm^2/s (Fig. 1.10). The solution of Fick's second law for the indicated initial concentration profile (where the N atoms do not cross the step) yields a Gaussian of the form $n(x,t) = N\Delta x/\sqrt{\pi D_x t} \cdot e^{-x^2/4D_x t}$, where $n(x, t)$ is the number of atoms contained in narrow stripes of width Δx parallel to the step and N is the total number of atoms. This profile at a certain time ($t = 118$ min) is shown in the inset of Fig. 1.10 with the full line being the Gaussian with $D_x = D_x^*$, that is, without fit parameter. Measurements at varying temperatures yield from $D_x^* = D_0 \cdot \exp(-E_{diff}^*/kT)$ the activation energy for surface diffusion $E_{diff}^* = 0.94 \pm 0.15$ eV and a prefactor $D_0 = 10^{-2}$ cm^2/s. As a

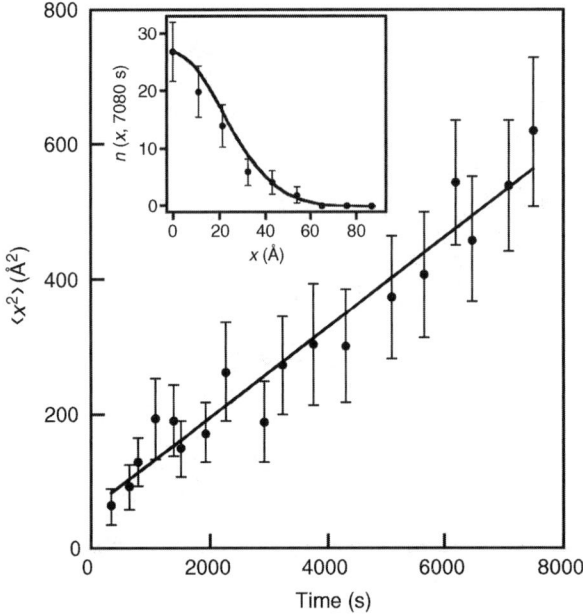

FIGURE 1.10. Diffusion of N atoms on a Ru(0 0 0 1) surface at 300K: mean square displacement of N atoms from a monoatomic step where they were created by dissociation of NO as a function of time. The inset shows the measured distribution of distances from the step after 2 h (points) and a Gaussian with adjusted parameters (full line) [40].

rule of thumb, E^*_{diff} is generally expected to be about 20% of the adsorption energy E_{ad}. In this case, $E_{ad} = 5.7\,\text{eV}$ [41], so the result for E^*_{diff} is consistent.

In general, except for such idealized cases as just described, the diffusion coefficient for adsorbates will be sensitively affected by the surface structure and the coverage. Nevertheless, data derived on larger scales by macroscopic techniques (such as photoemission electron microscopy (PEEM) [42]) will be of relevance for modeling surface reactions on these scales.

A special situation is found with chemisorbed hydrogen atoms where quantum effects come into play. The activation barrier for surface diffusion as sketched in Fig. 1.9b can become surmounted not only by thermal activation, but also (with low probability) by tunneling that manifests itself by the independence of temperature at low T [37]. Inspection of Fig. 1.9 indicates that a particle can reach a neighboring potential minimum by tunneling, where the probability depends on the mass of the particle and the height of the adjacent potential well. The latter is also affected by the zero point energy, so (apart from the mass difference) between H and D a pronounced isotope effect comes into play.

At higher temperatures, still another effect comes into play: vibrational excitation of adsorbed H atoms will cause transfer into delocalized band states. This effect was first predicted theoretically for H atoms adsorbed on Ni surfaces [43,44] and demonstrated by a combination of theory with vibrational spectroscopy for the system H/Pt(1 1 1) [45,46]. For not extremely low temperatures, the low-lying vibrational excitations will cause the H atoms to become delocalized (like the conduction electrons in a metal), and thus diffusion obtains a different meaning.

References

1. K. Honkala, A. Hellman, I. N. Remedakis, A. Logadottir, A. Carlsson, S. Dahl, C. H. Christensen, and J. K. Nørskov, *Science* **307** (2005) 555.
2. M. Haruta, *Chem. Phys. Chem.* **8** (2007) 1911, with references to related work.
3. I. Langmuir, *Trans. Faraday Soc.* **17** (1922) 607.
4. M. Boudart, *Catal. Lett.* **65** (2000) 1.
5. J. W. Niemantsverdriet, *Spectroscopy in Catalysis*, VCH, Weinheim, 1993.
6. J. C. Vickerman, *Surface Analysis: The Principal Techniques*, Wiley, 1997.
7. G. A. Somorjai, *Chemistry in Two Dimensions: Surfaces*, Cornell University Press, 1981.
8. K. W. Kolasinski, *Surface Science*, Wiley, 2002.
9. G. Ertl and J. Küppers, *Low Energy Electrons and Surface Chemistry*, 2nd ed., VCH, Weinheim, 1985.
10. D. P. Woodruff and T. A. Delchar, *Modern Techniques of Surface Science*, Cambridge University Press, 1994.
11. C. Feldman and J. W. Mayer, *Fundamentals of Surface and Thin Film Analysis*, North Holland, 1986.

12. (a) F. Besenbacher, Scanning tunneling microscopy studies of metal surfaces, *Rep. Prog. Phys.* **59** (1996) 1737; (b) F. Besenbacher, J. V. Lauritsen, and R. T. Wang, Scanning probe methods, in: *Handbook of Heterogeneous Catalysis*, Vol. **2** (eds. G. Ertl, H. Knözinger, F. Schüth, and J. Weitkamp), Wiley, 2008, p. 833.
13. F. Haber, *Z. Elektrochem.* **20** (1914) 521.
14. (a) I. Langmuir, *Phys. Rev. B* **8** (1916) 149; (b) I. Langmuir, *J. Am. Chem. Soc.* **40** (1918) 1361.
15. N. Al-Sarraf, J. T. Stuckless, C. E. Wartnaby, and D. A. King, *Surf. Sci.* **283** (1993) 427.
16. J. T. Stuckless, N. A. Frei, and C. T. Campbell, *Rev. Sci. Instrum.* **62** (1998) 2427.
17. G. Ertl and J. Koch, *Z. Naturforsch.* **25** (1970) 1906.
18. R. Hoffmann, *Acc. Chem. Res.* **4** (1971) 1.
19. S. Renisch, R. Schuster, J. Wintterlin, and G. Ertl, *Phys. Rev. Lett.* **82** (1999) 3839.
20. G. Ehrlich, *Adv. Catal.* **14** (1963) 255.
21. P. A. Redhead, *Vacuum* **12** (1962) 203.
22. D. A. King, *Surf. Sci.* **47** (1975) 384.
23. J. E. Lennard-Jones, *Trans. Faraday Soc.* **28** (1932) 333.
24. K. Christmann, O. Schober, G. Ertl, and M. Neumann, *J. Chem. Phys.* **60** (1974) 4528.
25. G. Ertl, in: *Metal Clusters in Catalysis* (eds. B. C. Gates, L. Guczi, and H. Knözinger), Elsevier, 1986, p. 589.
26. J. Wang, C. Fan, Q. Sun, K. Reuter, K. Jacobi, M. Scheffler, and G. Ertl, *Angew. Chem.* **42** (2003) 2151.
27. G. J. Kubas, *Acc. Chem. Res.* **21** (1988) 120.
28. P. Hohenberg and W. Kohn, *Phys. Rev.* **136** (1964) 864.
29. (a) B. Hammer and J. K. Nørskov, *Adv. Catal.* **45** (2006) 71; (b) A. Nilsson and L. G. M. Petterson, in: *Chemical Bonding at Surfaces and Interfaces* (eds. A. Nilsson, L. G. M. Petterson, and J. K. Norskov), Elsevier, 2008, p. 57.
30. P. J. Feibelman, B. Hammer, J. K. Nørskov, F. Wagner, M. Scheffler, R. Stumpf, R. Watwe, and J. Dumesic, *J. Phys. Chem. B* **105** (2001) 4018.
31. I. Langmuir, *J. Am. Chem. Soc.* **38** (1916) 2221; **40** (1918) 1361.
32. P. J. Kisliuk, *J. Phys. Chem. Solids* **3** (1957) 95; **5** (1958) 78.
33. D. A. King and M. G. Wells, *Surf. Sci.* **29** (1972) 454.
34. T. Engel, *J. Chem. Phys.* **69** (1978) 373.

35. T. E. Madey and J. T. Yates, *Surf. Sci.* **63** (1977) 203.
36. M. W. Roberts and C. S. McKee, *Chemistry of the Metal–Gas Interface*, Clarendon Press, Oxford, 1978.
37. R. Gomer, *Rep. Prog. Phys.* **53** (1990) 917.
38. G. Ehrlich, *Appl. Phys. A* **55** (1992) 403.
39. E. Ganz, S. K. Theiss, I.-S. Hwang, and J. Golovchenko, *Phys. Rev. Lett.* **68** (1992) 1567.
40. T. Zambelli, J. Trost, J. Wintterlin, and G. Ertl, *Phys. Rev. Lett.* **76** (1996) 795.
41. H. Shi, K. Jacobi, and G. Ertl, *J. Chem. Phys.* **99** (1993) 9248.
42. H. H. Rotermund, *J. Electron Spectrosc.* **98–99** (1999) 41.
43. K. Christmann, R. J. Behm, G. Ertl, M. A. Van Hove, and W. H. Weinberg, *J. Chem. Phys.* **70** (1979) 4168.
44. M. J. Puska, R. M. Niemenen, M. Manninen, B. Chakraborty, S. Holloway, and J. K. Nørskov, *Phys. Rev. Lett.* **51** (1983) 1081.
45. S. C. Badescu, P. Salo, T. Ala-Nissila, S. C. Ying, K. Jacobi, Y. Wang, K. Bedürftig, and G. Ertl, *Phys. Rev. Lett.* **88** (2002) 136181.
46. S. C. Badescu, K. Jacobi, Y. Wang, K. Bedürftig, G. Ertl, P. Salo, T. Ala-Nissila, and S. C. Ying, *Phys. Rev. B* **58** (2003) 205401.

CHAPTER 2

SURFACE STRUCTURE AND REACTIVITY

2.1. INFLUENCE OF THE SURFACE STRUCTURE ON REACTIVITY

Figure 2.1 shows the three most densely packed planes of fcc crystals. It is quite plausible that the varying coordination of the atoms in these surfaces will give rise to different chemisorption properties and hence reactivities. As a general rule, the less coordinatively saturated the surface atoms are the stronger the bonding will be, and as a consequence, the activation energy for dissociative chemisorption will show the opposite trend and the sticking coefficient will vary accordingly. This is, for example, demonstrated by the dissociative chemisorption of N_2 on different single-crystal surfaces of bcc Fe, as illustrated in Fig. 2.2 [1]. This difference will be of relevance for the catalytic formation of NH_3, as will be discussed later.

Even the best single-crystal surface cannot be perfect and will rather exhibit defects, mostly steps, and dislocations, as is evident from Fig. 2.3. In catalysis, monoatomic steps often play the role of "active sites" [2] with enhanced reactivity. An example is shown in Fig. 2.4 for the interaction of NO with a Ru(0001) surface

---

Reactions at Solid Surfaces. By Gerhard Ertl
Copyright © 2009 John Wiley & Sons, Inc.

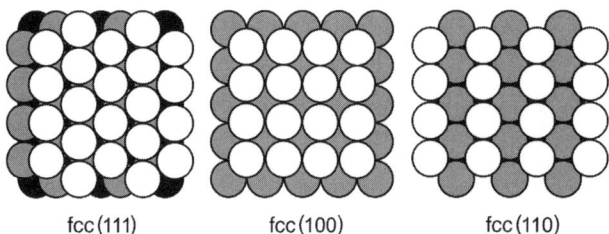

FIGURE 2.1. Ball models for the three most densely packed planes of fcc crystals.

exhibiting a monoatomic step (dark line) [3]. Six minutes after exposure to a small quantity of gaseous NO, the STM image exhibits dark triangular spots concentrated along the step and arising from the adsorbed N_{ad} species, while O_{ad} is much more mobile and diffuses rapidly away from the step and becomes discernible as weak streaks. After 120 min also the N_{ad} atoms have spread more across the terraces. Obviously, dissociation occurs preferentially at the step. This conclusion is supported by density functional theory (DFT) calculations [4], whereafter

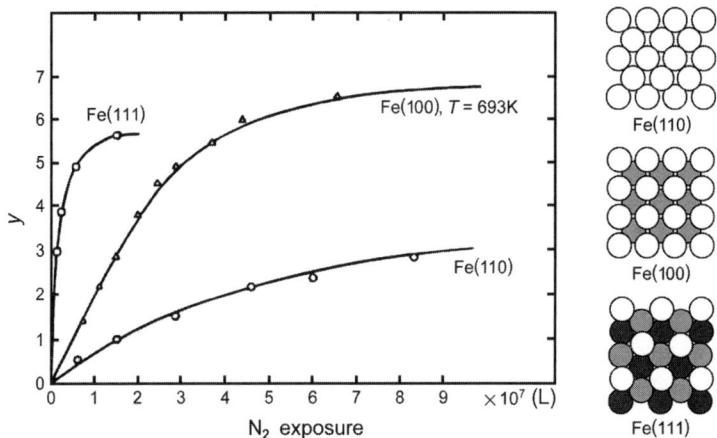

FIGURE 2.2. Variation of the relative coverage with N atoms, y, on different Fe single-crystal surfaces with N_2 exposure at 693K [1].

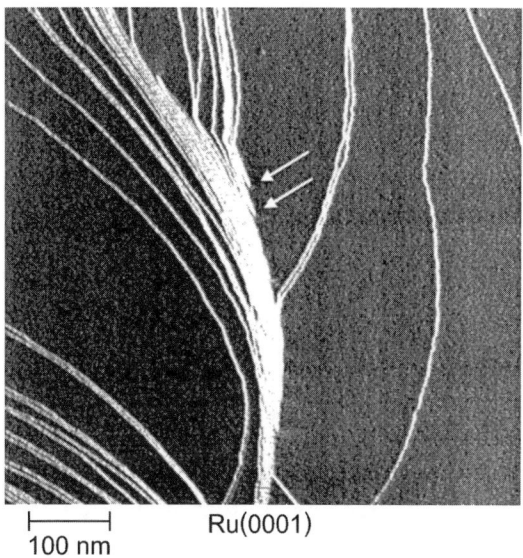

FIGURE 2.3. A Ru(0 0 0 1) surface exhibiting monoatomic steps and dislocations (arrows).

the activation energy for dissociation of an adsorbed NO molecule is 1.28 eV on the flat terrace, but only 0.15 eV at a step site. Interestingly, the actual density of steps is not decisive for the overall sticking coefficient (as long as steps are not too far from

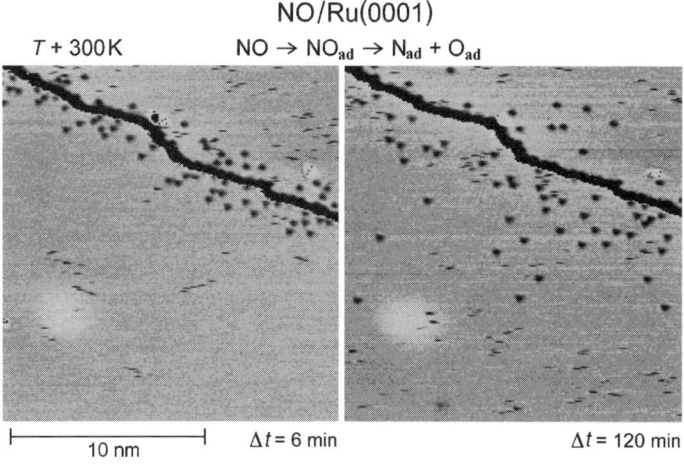

FIGURE 2.4. STM images from NO interacting with a Ru(0 0 0 1) surface exhibiting a monoatomic step [3].

each other): The terraces act as sinks for NO adsorption, and these particles then diffuse to the next step where they dissociate. Therefore, a larger terrace width compensates a lower step density.

In a catalytic reaction, all steps do not equally depend on the surface structure. So, for example, the rate of simple desorption processes is often not markedly affected by the structure of the surface. In catalysis, therefore, reactions are classified into "structure sensitive" and "structure insensitive" [5], usually on the basis of the variation of reactivity with particle size. As an example, the electrocatalytic oxygen reduction at platinum (which is of importance for fuel cells) will be mentioned, where a decrease of specific activity with increasing particle size was reported [6,7]. In a theoretical analysis [8], the kinetics was treated on the (1 1 1), (1 0 0), and (2 1 1) facets of several transition metals, and the results were combined with simple models for the geometries of catalytic nanoparticles. Thus, the experimentally observed trend could be well reproduced.

2.2. Growth of Two-Dimensional Phases

Figure 2.5a shows a snapshot from a Ru(0 0 0 1) surface with a small coverage of adsorbed O atoms at 300K. The O atoms are randomly distributed and move around like in a Brownian motion with a mean residence time (at 300K) of 60 ms at a certain adsorption site. However, due to the weak attraction between two adatoms with a minimum at a distance of $2a_0$ (a_0 = lattice constant of the substrate), at higher coverages a separation into two phases, namely, a quasi-gaseous and a quasi-crystalline phase, takes place (Fig. 2.5b) [9]. Under present conditions, the two phases are in equilibrium with each other, a situation that is rationalized by the phase diagram depicted in Fig. 2.6a. In our case, the horizontal scale (composition) denotes the concentration of occupied sites (i.e., overall coverage θ). As long as θ is small, we

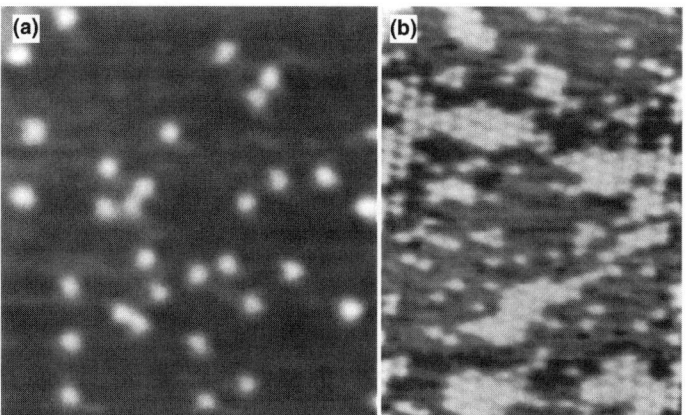

FIGURE 2.5. STM snapshots from O atoms adsorbed on a Ru(0 0 0 1) surface at 300K [9]: (a) at very low coverage; (b) at higher coverage.

are on the left side of the full line and in the region of a single phase (Fig. 2.5a). As soon as this full line is crossed, phase separation into a diluted (i.e., gaseous) and a more dense (i.e., crystalline) phase takes place, as demonstrated in Fig. 2.5b. Upon addition of more particles to the dilute phase, the new (dense) phase is formed. However, this process does not occur spontaneously but proceeds through nucleation: In a simple (continuum) picture, the free energy not only decreases proportional to r^2 and the area A occupied by the new phase but also increases in

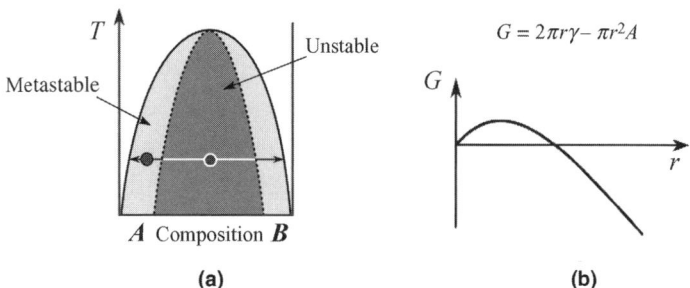

FIGURE 2.6. (a) Phase diagram for a binary system. (b) Variation of the free energy G of a nucleus with its radius r.

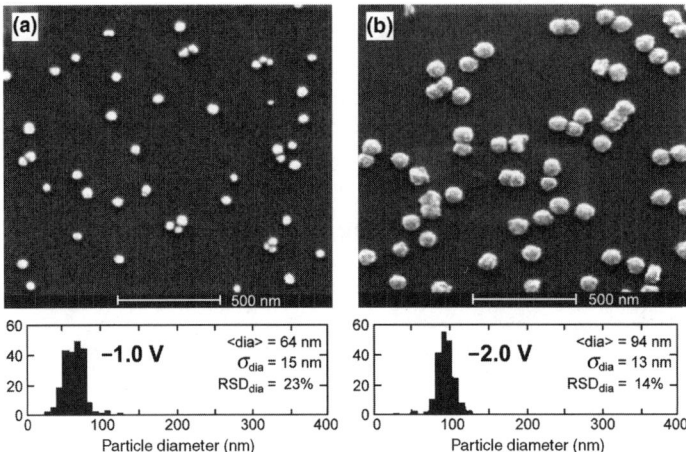

FIGURE 2.7. Formation of Ni nanoparticles with controlled size distribution by an electrochemical technique [10].

proportion to r due to the two-dimensional surface tension. The resulting curve is shown in Fig. 2.6b that passes with increasing r through a maximum, marking the critical size of the nucleus. This is the mechanism underlying the formation of nanoparticles whose average size can also be controlled in this way, as illustrated in Fig. 2.7. Further addition of material then leads to a competition between the rates of nucleation of new particles R_{nucl} and the growth of already existing ones R_{growth}, as depicted schematically in Fig. 2.8, and eventually the larger ones may grow on the expenses of smaller ones (Ostwald ripening) due to the further reduction of the surface tension term and lowering of the free energy. The final equilibrium state is frequently not reached because of the kinetic limitations, and thus metastable structures with phase boundaries and so on are formed. This is the general mechanism underlying the phenomenon called self-assembly. (Note that this is a situation for a thermodynamically closed system that tends to reach equilibrium, in contrast to open systems far from equilibrium where structure formation is denoted as (true) self-organization and will be discussed in detail later.)

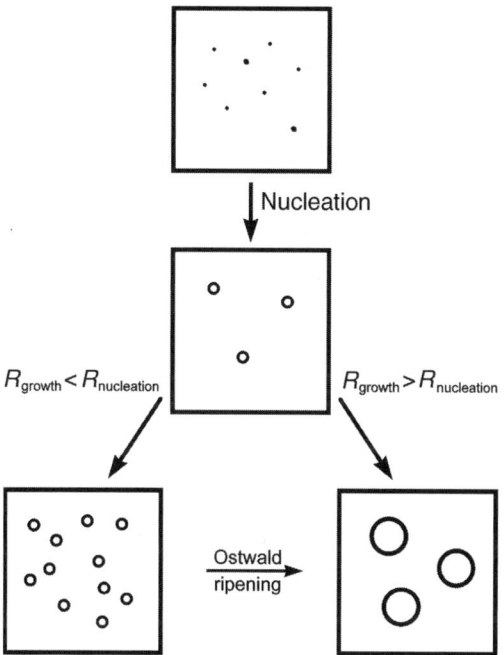

FIGURE 2.8. Principle of nucleation and growth processes.

Returning to Fig. 2.6a, the further we are inside the region marked by the solid line, the smaller will be the free energy for nucleation, and beyond the dashed line, it disappears completely and causes spontaneous phase separation called spinodal decomposition. The resulting situation is illustrated in Fig. 2.9 representing the result of a computer simulation in which a random distribution of particles with $\theta = 0.5$ and weak attractive interactions of $-4\,\text{kT}$ is allowed to start to diffuse at $t = 0$. The result is a labyrinthine pattern with characteristic nanometer length scale that is continuously growing with time as predicted by theory [11]. This new type of phase separation can only be observed if a point in the unstable region can be reached rapidly enough to suppress the ordinary nucleation–growth mechanism.

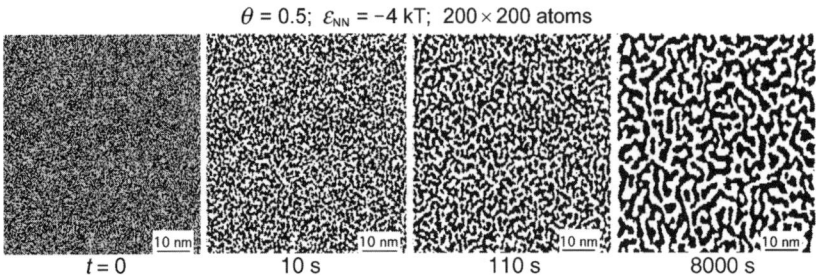

FIGURE 2.9. Computer simulation showing the formation of labyrinthine configurations following spinodal decomposition.

Figure 2.6a shows that this can be achieved by rapid quenching as, for example, with binary alloys [12]. However, this possibility fails with the formation of surface nanostructures, and hence practically (with the exception of polymers where diffusion is slow enough) no reports for spinodal structures exist. This problem could be overcome in a study in which about 50% of the atoms of the topmost layer of a Au(1 1 1) surface were removed within 20 µs by an electrochemical pulse techniques [13]. The result is shown in Fig. 2.10. In the STM image of Fig. 2.10a, the dark arrow

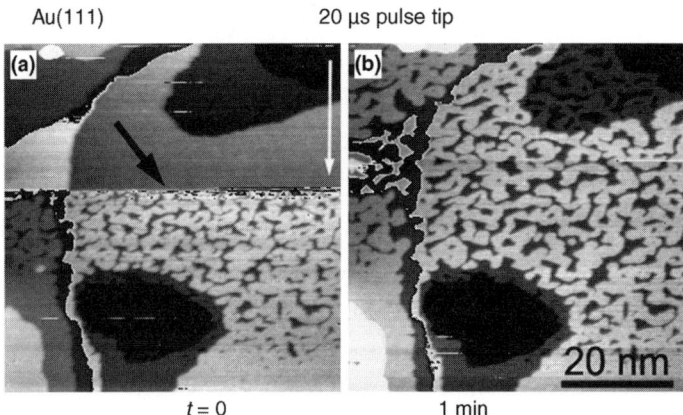

FIGURE 2.10. Formation of spinodal structures in the topmost monolayer of a Au(1 1 1) surface [13].

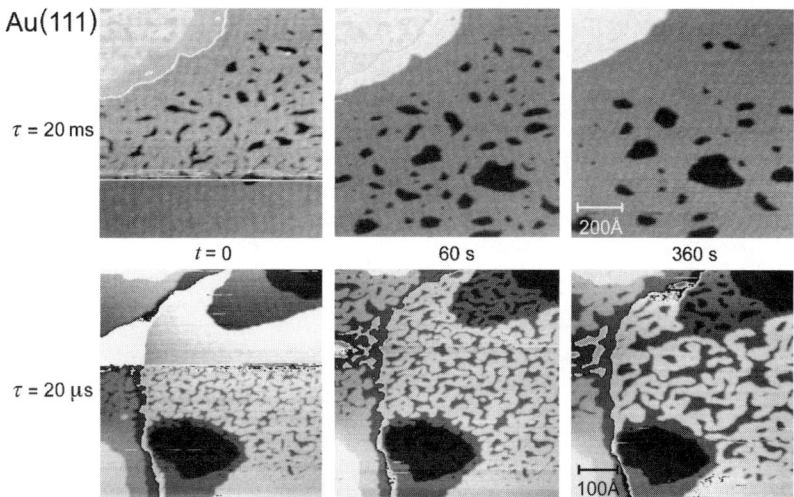

FIGURE 2.11. Patterns formed on a Au(1 1 1) surface after removal of half a monolayer of Au atoms within 20 ms (top) and 20 μs (bottom).

marks the point during the motion of the STM tip at which the desorption pulse was applied, and Fig. 2.10b shows the fully developed spinodal pattern as suggested by the simulation of Fig. 2.9. Figure 2.11 compares the situations resulting from slow (20 ms) desorption (i.e., nucleation and growth) in the top row with rapid (20 μs) creation (bottom) where the spinodal structure continuously exhibits Ostwald ripening.

2.3. Electrochemical Modification of Surface Structure

Figure 2.12 compares the energetics of a thermal reaction with that for a process associated with charge transfer across the interface. The latter type of reactions occur in electrochemical cells where a potential is applied between a working electrode and a counter electrode (Fig. 2.13). At the working electrode, the voltage U drops across the electrochemical double layer that has usually a thickness of only a few atomic layers, and the resulting high electric

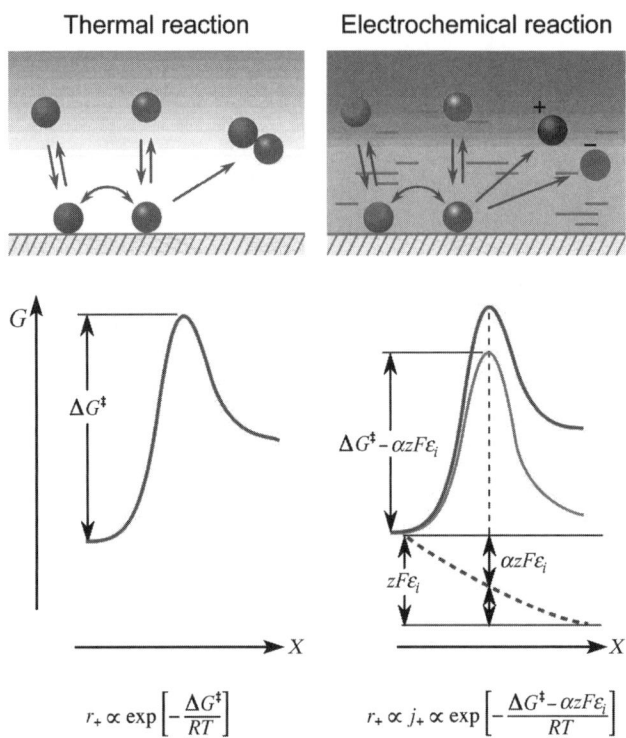

FIGURE 2.12. Energetics and kinetics of a thermal and an electrochemical reaction. (See color insert.)

field strength modifies the energy, for example, for removal of a surface atom via $A_s \rightarrow A^+ + e$. This was the basis for the creation of the spinodal structures described in the preceding section by applying a short voltage pulse between the gold surface and the nearby STM tip.

The lateral extension of the surface area whose structure can be affected in this way will be confined to the vicinity of the tip. The electrochemical fabrication of microstructures by approaching a tool to the surface is, however, limited by the nonuniform current distribution in this case, as illustrated in Fig. 2.14, and the resulting structure is rather disappointing. This problem may be overcome by applying short voltage pulses instead of a DC voltage, as shown in Fig. 2.15 [14]. If initiation of electrochemical dissolution requires

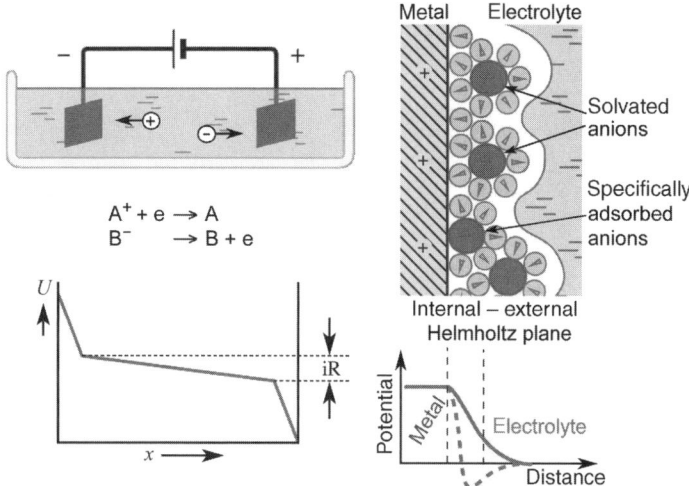

FIGURE 2.13. Potential distribution in an electrochemical cell (left) and structure and potential of the double layer (right). (See color insert.)

a minimum voltage at the working electrode U_{reac}, the time constant τ for reaching this voltage will be determined by the double layer capacity C_{DL} multiplied with the electrolyte resistance along the current path. If the pulse is short enough, this goal

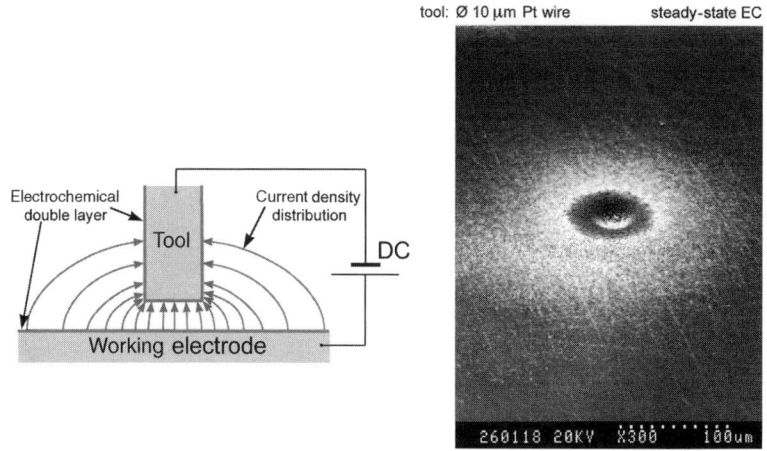

FIGURE 2.14. Electrochemical surface machining: current density distribution (left) and resulting hole (right). (See color insert.)

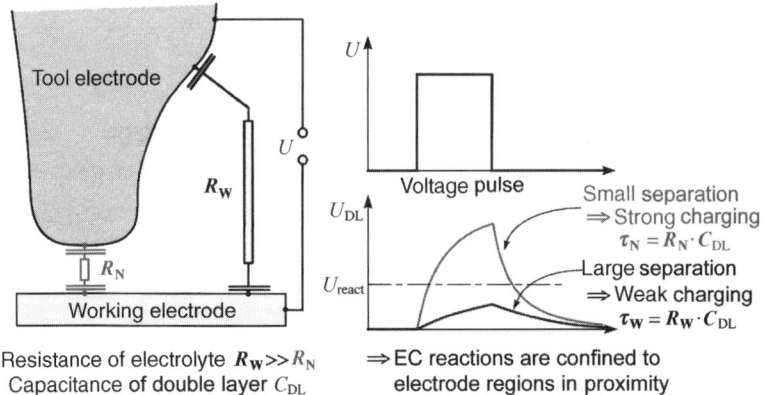

FIGURE 2.15. Principle of electrochemical micromachining by short voltage pulses [14]. (See color insert.)

will be reached only for regions in proximity to the tool electrode where $R_N \gg R_W$, with R_W being the resistance along longer current paths, that is, for regions farther away. The successful application of this principle is demonstrated in Fig. 2.16 that shows the creation of a microstructure in a Cu surface by applying

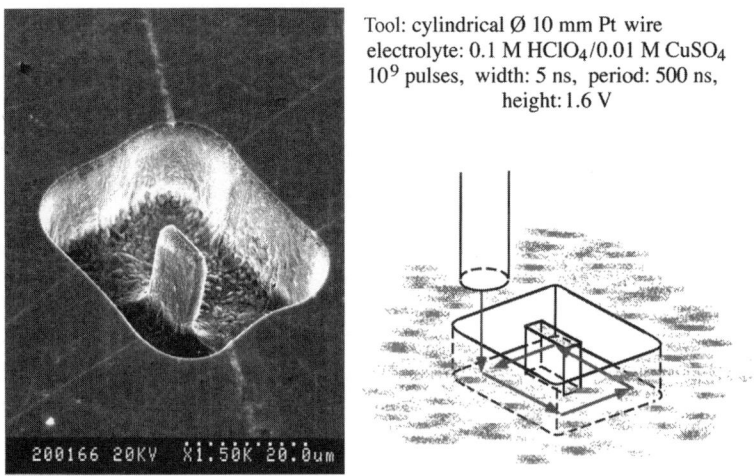

FIGURE 2.16. Electrochemical creation of a microstructure on a Cu surface [14]. (See color insert.)

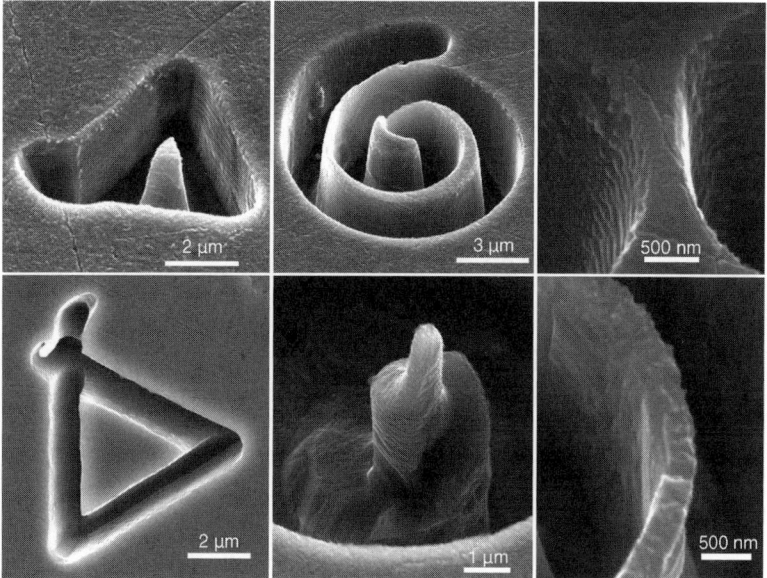

FIGURE 2.17. A series of microstructures on stainless steel fabricated by the electrochemical micromachining technique [15].

voltage pulses of 1.6 V and a width of 50 ns. Quite complicated structures can be fabricated in this way [15] such as shown, for example, in Fig. 2.17 for which resolutions far below 1 μm can be reached.

2.4. SURFACE RECONSTRUCTION AND TRANSFORMATION

The absence of part of the nearest neighbors that are present in the bulk no doubt affects the bonding as well as the structure of the atoms in the topmost layer. Even if their lateral configuration is identical to that of the corresponding bulk plane, the spacing between the first and the second layer will typically be slightly reduced, while that between the second and the third layer will be somewhat expended (relaxation). As outlined in Section 1.2, the energy of the adsorption bond is comparable to that between the surface atoms and hence adsorption will affect the surface structures. This effect was already realized by Langmuir [16]: "The atoms in the surface of a crystal must tend to arrange themselves so that the total energy will be a minimum. In general, this will involve a shifting of the positions of the atoms with respect to each other."

An example for the modification of the interlayer spacing by adsorption is shown in Fig. 2.18 for the system H/Ni(1 1 0) at $\theta = 1$ [17]. The adsorbed H atoms are located in threefold sites formed by atoms from the first and second Ni layers, where the spacing between the first and the second layer is compressed to 1.18 Å and that between the second and the third layer is expanded to 1.32 Å, while the bulk value is 1.25 Å.

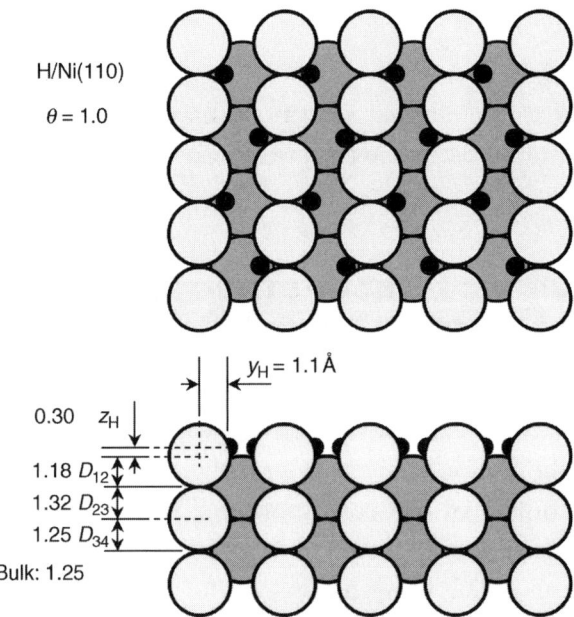

FIGURE 2.18. The structure of the Ni(1 1 0) surface with adsorbed H atoms at $\Theta = 1$ [17] (distances in Å).

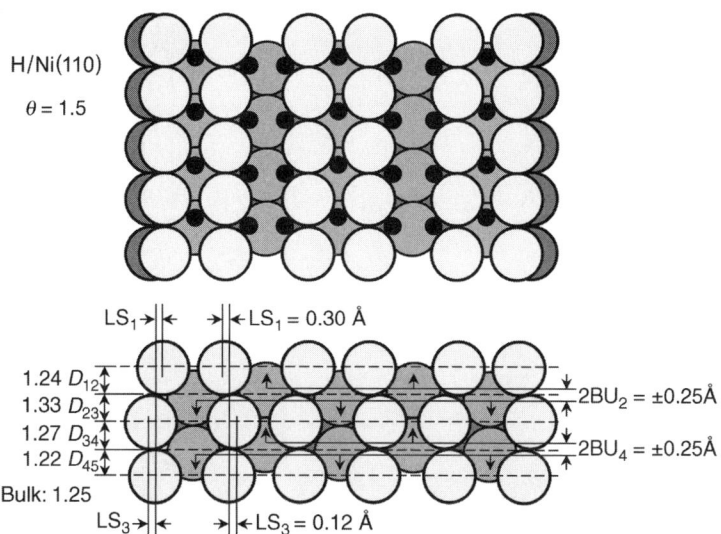

FIGURE 2.19. The structure of the Ni(1 1 0) surface with adsorbed H atoms at $\Theta = 1.5$ (distances in Å).

Figure 2.18 reveals that the threefold adsorption sites are not yet completely occupied, suggesting the possibility to increase the coverage even beyond $\theta = 1$. This is indeed the case and saturation is reached only at $\theta = 1.5$, for which the associated 1×2 structure is reproduced in Fig. 2.19. The H atoms are no longer adsorbed in identical sites, and rows of Ni atoms in the topmost layer have been shifted together by 0.6 Å to allow occupation of opposite threefold adsorption sites. As a consequence, the structures of even deeper layers are affected. This is a case of *displacive reconstruction* in which the symmetry of the surface unit cell is affected without a change in the atomic density.

True reconstruction implies a variation in the atomic density in the topmost layer of the substrate if compared with the corresponding bulk plane. Such a situation is found not only with numerous clean semiconductor surfaces but also with the (1 0 0) and (1 1 0) surfaces of the fcc 5d metals [18].

As an example, the situation for Pt(1 0 0) will be illustrated. The stable clean surface exhibits a quasi-hexagonal ("hex" or 5×20)

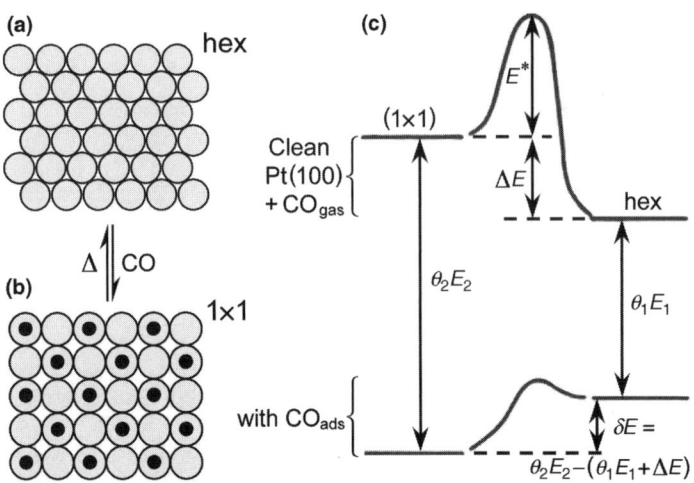

FIGURE 2.20. Lifting of the reconstruction of the hex Pt(1 0 0) surface by CO adsorption: (a) structure of the clean hex surface; (b) c2 × 2 structure of adsorbed CO after transformation of the substrate into its 1 × 1 structure; (c) energetics of this structural transformation process.

configuration of the atoms in the topmost layer [19–21], as depicted in Fig. 2.20a. The heat of CO adsorption is larger on the nonreconstructed 1 × 1 phase (Fig. 2.20b), so that the reconstruction is lifted. The transformation is initiated by a small CO coverage ($\Theta_{CO} \geq 0.05$) through a nucleation and island growth mechanism, until at $\Theta_{CO} = 0.5$ the whole surface has been converted into the 1 × 1 phase covered with a c2 × 2 CO overlayer [22,23]. The energetics of this transformation process are illustrated in Fig. 2.20c: Although the clean 1 × 1 phase is higher in energy than the clean hex phase, it is metastable and its thermal transformation is activated by $E^* \sim 20\text{--}25$ kcal/mol [24]. The heat of CO adsorption, on the other hand, is substantially smaller on the hex phase, the difference being the driving force for the phase transformation [23]. The first-order character of this phase transition becomes evident from the pronounced hysteresis effect if a Pt(1 0 0) surface is continuously heated up and subsequently cooled down in a constant CO atmosphere [25]. As shown in Fig. 2.21b, the CO coverage continuously decreases until

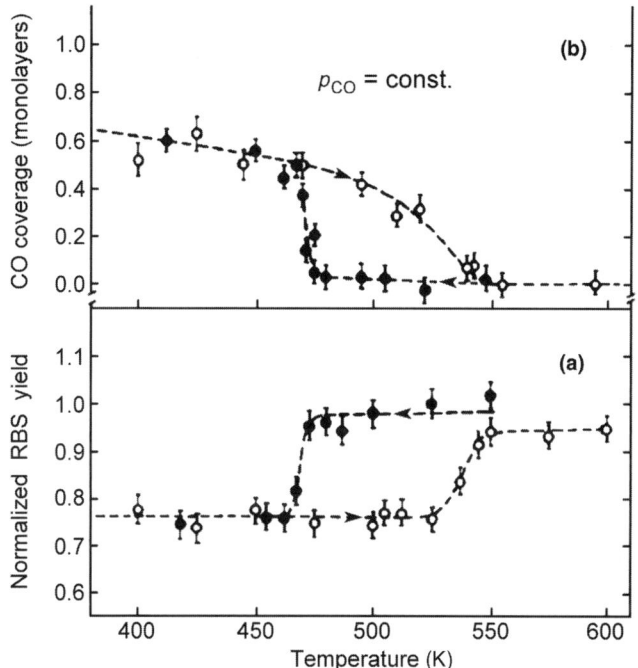

FIGURE 2.21. Hysteresis effects in the CO-induced hex → 1 × 1 structural transformation of Pt(1 0 0) [25]: (a) structural parameter reflected by the Rutherford backscattering yield; (b) CO coverage.

transformation into the hex phase (as reflected by the structural data plotted in Fig. 2.21a and derived from Rutherford backscattering (RBS) experiments) is completed. Upon cooling, the surface now remains in the hex structure down to about 470K when enough adsorbed CO has accumulated to cause rapid transformation into the 1×1 phase accompanied by a strong increase of the CO coverage.

The difference in the atomic density in the topmost layer becomes even more pronounced with the Pt(1 1 0) surface, which in the clean stable state exhibits a 1×2 missing row structure, while CO adsorption causes transformation in the bulk-like 1×1 structure that contains twice as many surface atoms (Fig. 2.22). In this case, at 300K, the transformation is initiated by homogeneous

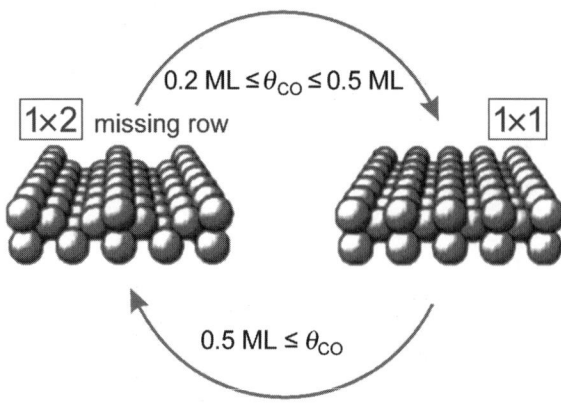

FIGURE 2.22. The CO-induced $1 \times 2 \rightarrow 1 \times 1$ structural transformation of the Pt(1 1 0) surface.

nucleation of small 1×1 patches exposed from the second layer. In this way, migration of surface Pt atoms is restricted to only a few lattice sites. At higher temperatures, on the other hand, lateral displacements of longer chains take place, and the enhanced surface mobility enables the formation of larger anisotropic 1×1 islands [26].

The CO-induced structural transformations of both the Pt(1 0 0) and Pt(1 1 0) surfaces will be of decisive importance for the occurrence of kinetic oscillations in CO oxidation, as will be discussed in detail in Chapter 7.

A quite interesting alternative for the formation of an adsorbate-induced surface reconstruction is offered by the O/Cu(1 1 0) surface. Upon adsorption of O atoms on the clean (nonreconstructed) 1×1 surface, a (2×1)-O–Cu(1 1 0) phase is formed whose structure is depicted in Fig. 2.23c. This new "missing row" structure is, however, formed by condensation of mobile chemisorbed O atoms with Cu adatoms evaporating from steps and diffusing across the terraces of the substrate surface. This nucleation process is illustrated by the STM snapshots in Fig. 2.23a and b leading to the structure of Fig. 2.23c that is more appropriately described as "added row" than "missing row" phase [27].

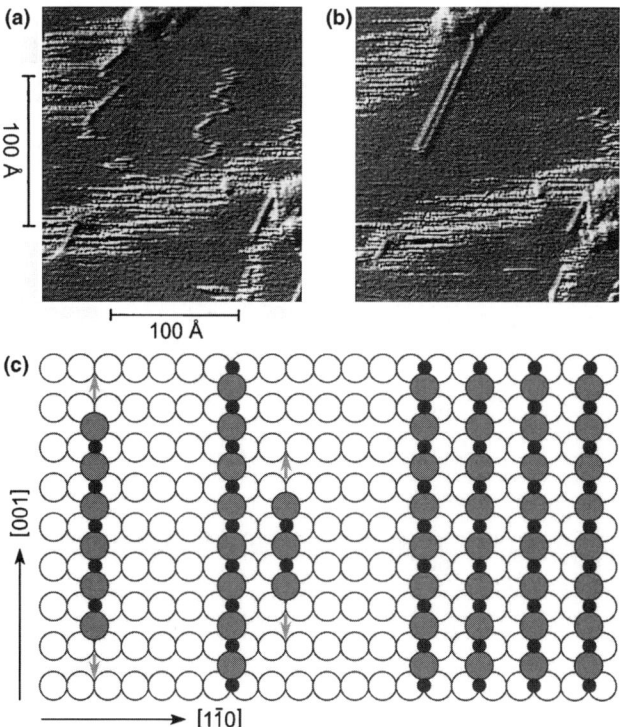

FIGURE 2.23. The formation of the (2 × 1)-O–Cu(1 1 0) "added row" structure [27]: (a and b) subsequent STM snapshots; (c) ball model illustrating the formation process.

Apart from chemisorption, the state of reconstruction of a surface may be affected by an electric potential across the surface/solution interface in an electrochemical system [28]. It was found that at electrode potentials positive to the potential of zero charge, the reconstruction is lifted and the surface changes to the bulk-truncated structure. Thus, the hex structure of a Au(1 0 0) electrode is lifted in 0.01 M $HClO_4$ solution at $E > 0.60$ V versus saturated calomel electrode, SCE, but already at $E > 0.27$ V versus SCE in 0.01 M H_2SO_4 solution, indicating the additional role of specific adsorption. These findings could recently be surprisingly well reproduced theoretically by a combination of DFT and thermodynamic considerations [30].

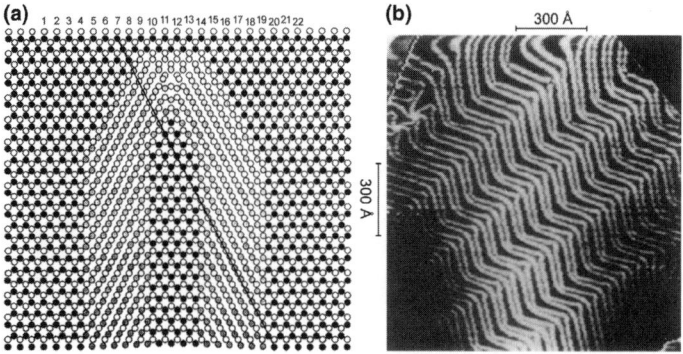

FIGURE 2.24. The reconstructed Au(1 1 1) surface: (a) model of the atomic structure of a U connection between two corrugation lines; (b) STM image from the zigzag configuration.

A very particular role is played by the clean Au(1 1 1) surface, which is the only most densely packed metal surface exhibiting reconstruction. The surface atoms are uniformly contracted along the $[1\bar{1}0]$ direction so that 23 surface atoms are located over 22 hollow sites of the second layer leading to parallel corrugation lines along the $[1\bar{1}2]$ direction. Individual corrugation lines, separating different stacking regions, cannot disappear, but are frequently terminated by U-shaped connections between the neighboring lines, as depicted in Fig. 2.24a [31].

In addition, a new long-range superstructure is created by the correlated periodic bending of the parallel corrugation lines by ±120° every 25 nm, so that rotational domains are arranged in a zigzag pattern, as evident from the STM image of Fig. 2.24b. Interactions on such large scale indicate their origin in the long-range elastic strain. The structure reflects the overall tendency to isotropic contraction, combining the locally favorable uniaxial contraction and an effective isotropic contraction on a larger scale.

A similar effect of strain-induced long-range order has also been observed with the "added row" 2 × 1-O–Cu(1 1 0) system discussed above [32]. On the basis of He diffraction and STM experiments, it was found that the anisotropic Cu–O islands, as depicted in Fig. 2.23, arrange themselves in a periodic super-grating with stripes along the [1 0 0] direction and their periodicity ranging between 6 and 14 nm, depending on O coverage and temperature. It was suggested that substrate-mediated elastic interactions are responsible for this effect. Later, it was shown that the long-range interactions (typically up to several nm) between (metallic) adsorbates may also be mediated through electronic surface states [33].

The tendency for surface restructuring by adsorption to minimize the free energy can proceed beyond the topmost layers and lead to the formation of new crystal planes (faceting). These facets are usually inclined against the original substrate plane and the initial stages of their formation can therefore be conveniently followed by low-energy electron diffraction (LEED). As an example, Fig. 2.25a shows the continuous splitting of the (0,1)-LEED

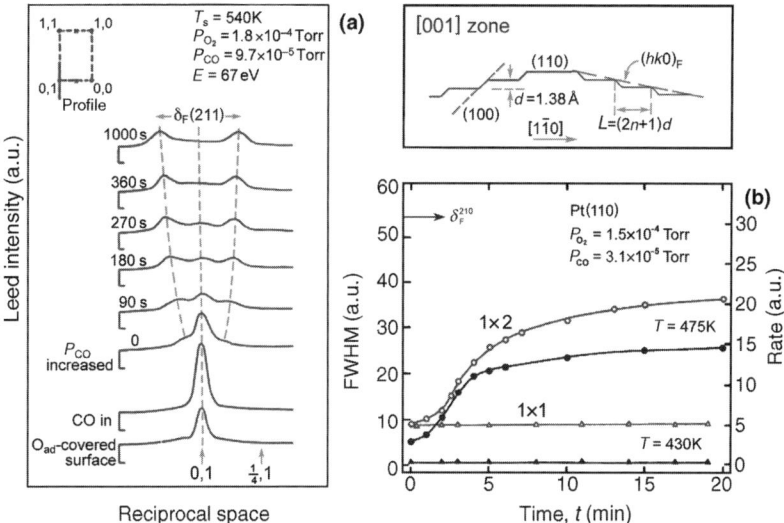

FIGURE 2.25. Facet formation in the course of catalytic CO oxidation on a Pt(1 1 0) surface [34]: (a) continuous splitting of the (0,1)-LEED beam and sketch of the facets; (b) variation of the beam splitting (open symbols) and the steady-state reaction rate (filled symbols) at two temperatures.

beam of a Pt(1 1 0) surface under the influence of CO oxidation under steady-state flow conditions, eventually leading to the formation of (2 1 1) facets [34]. This type of reconstruction may be accompanied by a continuous change of the catalytic activity, that is, the reaction "digs its own bed." Such an effect had already been observed by Langmuir [35] in CO oxidation on a Pt wire, who concluded "...indicating that the filament was undergoing a progressive change in the direction of becoming a better catalyst" and speculated that "There is good evidence that the effect is caused by changes in the structure of the surface itself, brought about by the reaction."

With our example, a continuous increase of the reaction rate is directly paralleled by the progress of faceting monitored by the splitting of the relevant LEED beam, as shown in Fig. 2.25b.

2.5. Subsurface Species and Compound Formation

Since the surface atoms have fewer next neighbors than those in the bulk, chemisorption is generally characterized by stronger bond formation and occurs prior to bulk compound formation. However, atoms may often be dissolved in the bulk and—depending on temperature—segregate to the surface where they may affect the reactivity. Such an example was found with the catalytic decomposition of NH_3 [36] as well as with dissociative hydrogen adsorption [37] on Ni surfaces. According to the previous reports, the reactivity should change upon crossing the Curie temperature (magnetocatalytic or Hedvall effect). It turned out, however, that this effect is caused by segregation of carbon from the bulk to the surface, which is affected by a slight variation of the Ni lattice constant at the Curie temperature. Rigorously clean surfaces do not exhibit such an effect. A nice example for the role of subsurface atoms on catalytic activity was recently reported for alkyne hydrogenation on palladium [38]. It was found that the population of subsurface sites with both hydrogen and carbon atoms governs the selectivity of catalytic hydrogenation on the surface.

With hydrogen, penetration of the adsorbed atoms below the surface is particularly pronounced in the case of Pd because of the possibility for the formation of bulk hydride phases [39]; however, it may also be relevant for hydrogen on Ni or oxygen on Pt [40].

The formation of bulk phases is most common in the case of oxygen. Transformation of the chemisorbed phase into oxide proceeds generally through a nucleation and growth mechanism. As an example, Fig. 2.26a shows an STM image from a Ru(0 0 0 1) surface that had been exposed to O_2 at elevated temperature [41]. The right part is still the Ru(0 0 0 1) surface covered by a 1×1 O adlayer, while the left part had been transformed into a thin

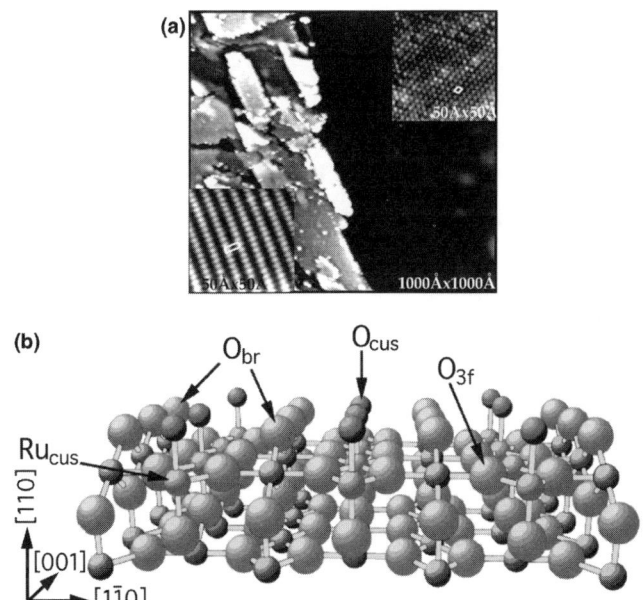

FIGURE 2.26. Structural transformation of the Ru(0 0 0 1) surface into a RuO_2(1 1 0) overlayer under the influence of oxygen: (a) STM image exhibiting both phases; (b) ball model of the RuO_2(1 1 0) surface. (See color insert.)

overlayer terminated by the (1 1 0) surface of RuO_2. Figure 2.26b shows a ball model of this surface in which Ru atoms (blue) are coordinated to bridging O atoms (O_{br}), while additional Ru atoms (red) are coordinatively unsaturated (Ru_{cus}) and may adsorb either additional O atoms (O_{cus}) or other species, thus becoming relevant for catalytic reactivity, as will be discussed in Chapter 6.

Another interesting and widely studied example is the oxidation of aluminum. STM studies with an Al(1 1 1) surface at room temperature [42] revealed that the initial sticking coefficient is rather small and leads to the formation of O adatoms that at higher coverages agglomerate to small 1×1 islands. Distinction between chemisorbed and oxide oxygen can be made on the basis of the shape of the Al^{3+} Auger spectra. Oxidation starts long before completion of the 1×1 adlayer at coverages around $\Theta_O = 0.2$. With increasing O_2 exposure, both phases grow simultaneously in coverage until no bare surface is available anymore. From this point onward, further oxygen uptake leads to conversion of O_{ad} areas to oxide up to full coverage with the latter. Oxide nucleation takes place at the interface of O_{ad} islands and bare surface until the surface is completely covered with a layer of small oxide particles of about 2 nm diameter. Finally, the whole surface is covered by a relatively smooth, amorphous layer of Al_2O_3.

At elevated temperatures ($\geq 530K$), the chemisorbed O atoms are more mobile and form larger 1×1 O_{ad} islands, but nucleation and formation of the oxide proceed along with similar mechanisms [43].

2.6. Epitaxy

Epitaxy denotes the growth of a second (bulk) phase on a single crystalline substrate and plays an important role, for example, in semiconductor technology. Several techniques are in use to grow such composite layers apart from simple evaporation. In molecular beam epitaxy (MBE), the second material is evaporated within a Knudsen cell as a thermal beam onto the single-crystal substrate. In chemical vapor deposition (CVD), molecular precursors are impinging onto the substrate from the gas phase where they decompose. Still another possibility consists in deposition by electrochemical reaction.

The growth of a crystalline phase B on top of another one (A) will generally be associated with a mismatch of the lattice constants, and this effect will in turn cause some strain in the overlayer. These strain effects will in turn be responsible for the growth mode in solid-on-solid epitaxy. The essential three growth modes are depicted schematically in Fig. 2.27.

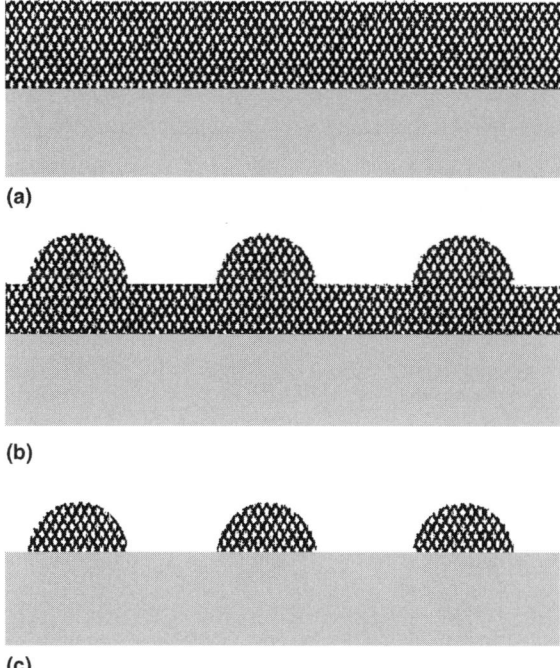

FIGURE 2.27. Equilibrium growth modes in epitaxy: (a) layer-by-layer (Frank–van der Merwe); (b) layer plus island (Stranski–Krastanov); (c) island growth (Volmer–Weber).

(a) *Layer-by-layer (Frank–van der Merwe) growth.* Phase B grows with the same lattice constant as A (pseudomorphic growth), where the lattice misfit leads to the buildup of elastic stress that at a critical thickness is released through the formation of dislocations. An example for this growth mode is offered by the electrochemical deposition of Cu on a Pt(1 0 0)-1 × 1 surface [45]. Cu grows layer-by-layer up to five atomic layers and then relaxes to the (1 0 0)-oriented Cu bulk structure leading to a typical Moiré pattern in STM with its periodicity determined by the Cu–Pt lattice misfit. The relatively large thickness of the pseudomorphic film that is under tensile stress has to be attributed to the counteracting compressive stress caused by adsorbed sulfate ions.

(b) *Layer plus island (Stranski–Krastanov) growth.* After the formation of a wetting interface, the overlayer does not continue to grow in the strained structure and instead forms three-dimensional islands. This growth mode is most frequently observed and governs, for example, the formation of the so-called quantum dots [46]. These are small structures (often with narrow size distribution) from, for example, III–V semiconductors on other semiconductor substrates that exhibit special optical properties and are the subject of intense current interest.

(c) *Three-dimensional island (Volmer–Weber) growth.* The lattice mismatch is too great so that from the beginning islands with the lattice of phase B are formed.

Often the resulting structure is not in thermodynamic equilibrium but is governed by kinetic effects, in particular by diffusion of the adsorbed atoms across the terraces and over the monoatomic steps [47].

References

1. F. Bozso, G. Ertl, M. Grunze, and M. Weiss, *J. Catal.* **49** (1977) 18; **50** (1977) 519.
2. H. S. Taylor, *Proc. R. Soc. A* **108** (1925) 105.
3. T. Zambelli, J. Wintterlin, J. Trost, and G. Ertl, *Science* **273** (1996) 1688.
4. B. Hammer, *Phys. Rev. Lett.* **83** (1999) 3681.
5. M. Boudart, *Adv. Catal.* **20** (1970) 153.
6. K. J. J. Mayrhofer, B. B. Blizanac, M. Arenz, W. R. Stamkenovic, P. N. Ross, and N. M. Markovic, *J. Phys. Chem. B* **109** (2005) 14433.
7. H. A. Gasteiger, S. S. Kocha, B. Sompalli, and F. T. Wagner, *Appl. Catal. B* **52** (2005) 9.
8. J. Greenley, J. Rossmeisl, A. Hellmann, and J. K. Nørskov, *Z. Phys. Chem.* **221** (2007) 1209.
9. J. Wintterlin, J. Trost, S. Renisch, R. Schuster, T. Zambelli, and G. Ertl, *Surf. Sci.* **394** (1997) 159.
10. M. P. Zach and R. M. Penner, *Adv. Mater.* **12** (2000) 878.
11. J. W. Cahn, *J. Chem. Phys.* **42** (1965) 93.
12. K. Oki, et al. *J. Phys. C* **7** (1977) 414.
13. R. Schuster, D. Thron, M. Binetti, X. Xia, and G. Ertl, *Phys. Rev. Lett.* **91** (2003) 066101.
14. R. Schuster, V. Kirchner, P. Allongue, and G. Ertl, *Science* **289** (2000) 98.
15. V. Kirchner, L. Cagnon, R. Schuster, and G. Ertl, *Appl. Phys. Lett.* **79** (2001) 1721.
16. I. Langmuir, *J. Am. Chem. Soc.* **38** (1916) 1221.
17. W. Reimer, V. Penka, M. Skottke, R. J. Behm, G. Ertl, and W. Moritz, *Surf. Sci.* **186** (1987) 45.
18. G. A. Somorjai, *Introduction to Surface Chemistry and Catalysis*, Wiley, 1994.
19. M. A. Van Hove, R. J. Koestner, P. C. Stair, J. P. Biberian, L. L. Kesmodel, I. Bartos, and G. A. Somorjai, *Surf. Sci.* **103** (1981) 189.
20. K. Heinz, E. Lang, K. Strauss, and K. Müller, *Surf. Sci.* **120** (1982) L401.
21. P. Heilmann, K. Heinz, and K. Müller, *Surf. Sci.* **83** (1979) 487.
22. R. J. Behm, P. A. Thiel, P. R. Norton, and G. Ertl, *J. Chem. Phys.* **78** (1983) 7437.

23. P. A. Thiel, R. J. Behm, P. R. Norton, and G. Ertl, *J. Chem. Phys.* **78** (1983) 7498.
24. P. R. Norton, J. A. Davies, D. K. Creber, C. W. Sitter, and T. E. Jackman, *Surf. Sci.* **108** (1981) 205.
25. T. E. Jackman, K. Griffiths, J. A. Davies, and P. R. Norton, *J. Chem. Phys.* **79** (1983) 3529.
26. T. Gritsch, D. Coulman, R. J. Behm, and G. Ertl, *Phys. Rev. Lett.* **63** (1989) 1086.
27. D. J. Coulman, J. Wintterlin, R. J. Behm, and G. Ertl, *Phys. Rev. Lett.* **64** (1990) 1761.
28. D. M. Kolb, *Prog. Surf. Sci.* **51** (1996) 109.
29. D. M. Kolb, *Surf. Sci.* **500** (2002) 722.
30. S. Venkatachalam, P. Kaghazchi, L. A. Kibler, D. M. Kolb, and T. Jacob, *Chem. Phys. Lett.* **455** (2008) 47.
31. J. V. Barth, H. Brune, G. Ertl, and R. J. Behm, *Phys. Rev. B* **42** (1990) 9307, and references therein to previous work.
32. K. Kern, H. Niehus, A. Schatz, P. Zeppenfeld, J. Goerge, and G. Comsa, *Phys. Rev. Lett.* **67** (1991) 855.
33. N. Knorr, H. Brune, M. Epple, A. Hirstein, M. A. Schneider, and K. Kern, *Phys. Rev. B* **65** (2002) 115420.
34. S. Ladas, R. Imbihl, and G. Ertl, *Surf. Sci.* **197** (1988) 153.
35. I. Langmuir, *Trans. Faraday Soc.* **17** (1921) 607, 621.
36. G. Ertl and J. Rüstig, *Surf. Sci.* **119** (1982) L314
37. H. Robota, W. Vielhaber, and G. Ertl, *Surf. Sci.* **136** (1984) 111.
38. D. Teschner, J. Borsodi, A. Wootsch, Z. Revay, M. Hävecker, A. Knop-Gericke, S. D. Jackson, and R. Schlögl, *Science* **320** (2008) 86.
39. H. Conrad, G. Ertl, and E. E. Latta, *Surf. Sci.* **41** (1974) 435.
40. A. von Oertzen, A. S. Mikhailov, H. H. Rotermund, and G. Ertl, *J. Phys. Chem. B* **102** (1998) 4966.
41. H. Over, Y. D. Kim, A. P. Seitsonen, S. Wendt, E. Lundgren, M. Schmid, P. Varga, A. Morgante, and G. Ertl, *Science* **287** (2000) 1474.
42. H. Brune, J. Wintterlin, J. Trost, G. Ertl, J. Wiechers, and R. J. Behm, *J. Chem. Phys.* **99** (1993) 2128.
43. J. Trost, H. Brune, J. Wintterlin, R. J. Behm, and G. Ertl, *J. Chem. Phys.* **108** (1998) 1740.
44. K. W. Kolasinski, *Surface Science*, Wiley, 2002, p. 258 ff.

45. A. M. Bittner, J. Wintterlin, and G. Ertl, *Surf. Sci.* **376** (1997) 267.
46. K. Jacobi, *Prog. Surf. Sci.* **71** (2003) 185.
47. G. Rosenfeld, B. Poelsema, and G. Comsa, in: *The Chemical Physics of Solid Surfaces and Heterogeneous Catalysis,* Vol. **8** (eds. D. A. King and D. P. Woodruff), Elsevier, 1997, p. 66.

CHAPTER 3

DYNAMICS OF MOLECULE/ SURFACE INTERACTIONS

3.1. INTRODUCTION

The progress of a chemical reaction comprises variation of the positions of the atomic nuclei along a multidimensional potential hyperface. The path with minimum energy with respect to the other degrees of freedom is called the reaction coordinate. If the potential along this coordinate exhibits a maximum, it is called the transition state and its difference to the initial state is denoted as activation energy E^*. From the partition functions of the initial and transition states, the activation entropy $\Delta S^{\#}$ is derived. Within the framework of transition state theory (TST) [1], the rate constant for the reaction is then given by

$$k_r = \frac{k_B T}{h} e^{\Delta S^{\#}/k_B} \cdot e^{-E^*/k_B T} \qquad (3.1)$$

where k_B is Boltzmann's constant and h is Planck's constant.

The basic idea of TST consists in the assumption that, at all stages along the reaction coordinate, thermal equilibrium is established, so the temperature T is the only essential (macroscopic) parameter. In the case of surface reactions, this requires

---

Reactions at Solid Surfaces. By Gerhard Ertl
Copyright © 2009 John Wiley & Sons, Inc.

that energy exchange between the various degrees of freedom of the particle interacting with the surface occurs much faster than the process of chemical transformation (which typically takes place on the timescale of picoseconds). If the substrate is a metal, typical relaxation times for excitations are about 10^{-15} s for electrons, 10^{-12} s for nuclei (phonons), and 10^{-8} s for photons. It is quite obvious that the long lifetime of photons prevents any true photochemical reactions (at metal surfaces) and the excitation is instead overtaken by the electrons. The dynamics of surface processes governed by electronic excitations will be discussed in Chapter 4. It will turn out that for reactions within the adsorbed phase, relaxation with the electrons and phonons will usually be fast enough to validate TST. However, if the particle is either approaching or leaving the surface, energy exchange will be restricted to a rather narrow range and hence the requirements for TST will often not be fulfilled. The associated effects are dealt with in the present chapter.

3.2. Scattering at Surfaces

The nonthermal character of molecule/surface interaction becomes evident from the angular distribution of particles from a monochromatic molecular beam scattered at a surface, as reproduced in Fig. 3.1 for a Pd(1 1 1) surface [2].

(a) He undergoes *direct elastic scattering* due to reflection at a practically hard wall without energy exchange. In experiments with very high-energy resolution, however, even excitation of single phonons may be discerned in such types of experiments [3,4]. Even the wave nature of atoms or molecules becomes evident by diffraction at the atomic grating of a single-crystal surface. Figure 3.2 shows angular distributions along the [0 0 1] azimuth of a Ni(1 1 0) surface from H_2 and D_2 beams, where apart from the

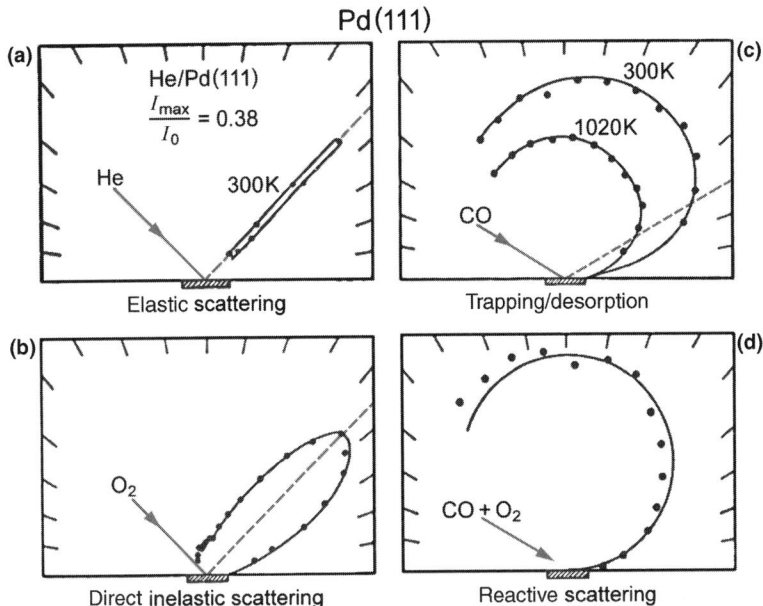

FIGURE 3.1. Polar plots of the angular distributions of the various scattering types at a Pd(1 1 1) surface [2].

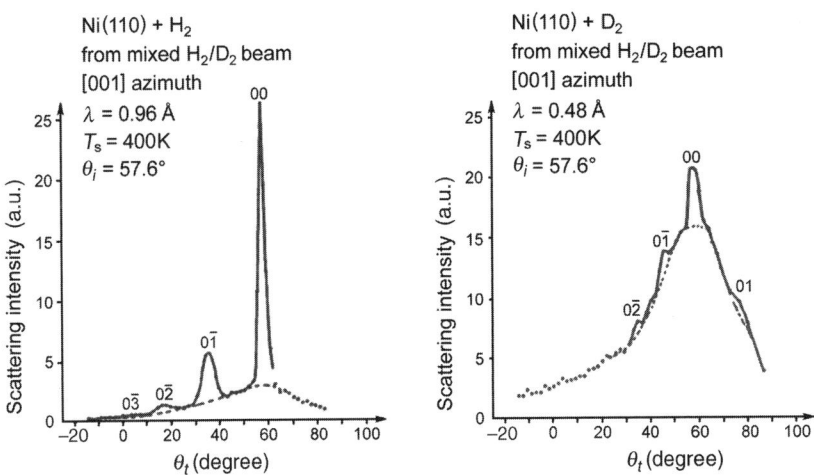

FIGURE 3.2. Angular distributions of H_2 and D_2 molecular beams scattered from a Ni(1 1 0) surface [5].

directly reflected 00 beam, additional peaks due to diffraction are discernible [5] since the de Broglie wavelengths of H_2 and D_2 are compatible with the Ni(1 1 0) lattice constant.

(b) Dotted lines in Fig. 3.2 mark a background arising from *direct inelastic scattering*. Here the particles exchange energy upon collision with the surface. These are not completely accommodated with the heat bath of the solid, but retain some of their "memory" of the initial momentum. The angular distributions in Fig. 3.2 are still peaked in the specular direction, but are broadened. In Fig. 3.2, the higher mass of D_2 if compared with H_2 is responsible for the enhanced contribution from this channel.

(c) A cosine angular distribution (Fig. 3.1c) indicates that the particles coming off the surface had undergone *trapping/desorption*, that is, they were fully thermally accommodated with the solid before desorbing after a mean surface residence time τ. The lower density recorded at higher (1020K) than at lower temperature (300K) arises from the fact that molecules coming off a hotter surface are also faster.

(d) *Reactive scattering* finally denotes the phenomenon that species other than those impinging come off the surface as a consequence of surface reaction. The angular distribution is then usually symmetric with respect to the surface normal, but not necessarily cosine as will be outlined below for the case of associative desorption.

3.3. Dissociative Adsorption

The Lennard–Jones potential diagram reproduced in Fig. 1.4 illustrates the progress of dissociative adsorption only in a simplified one-dimensional manner. In reality, this process is governed by a multidimensional potential involving all the nuclear coordinates. Figure 3.3c schematically shows the energetics of dissociative adsorption in the form of contour plots as a function of two coordinates, the distance x of the diatomic molecule from the surface and the separation y between the two atoms.

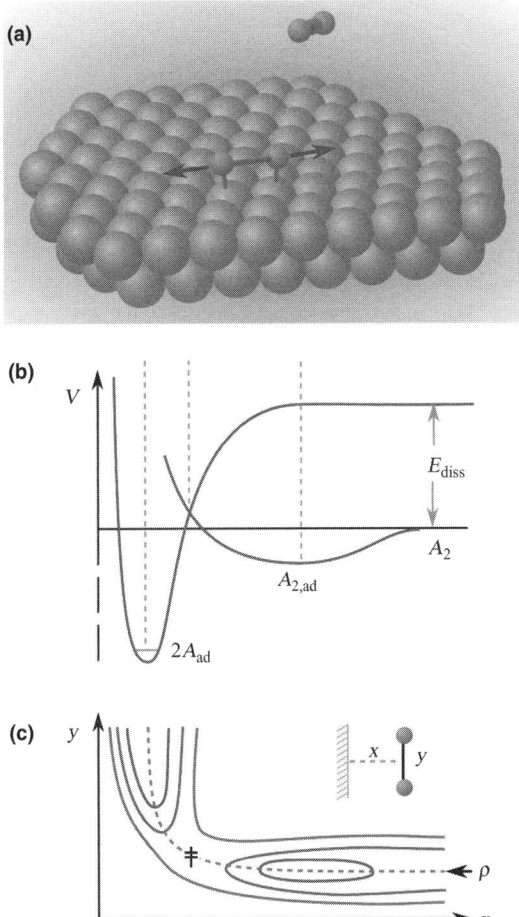

FIGURE 3.3. Energetics of dissociative adsorption of a diatomic molecule. (a) Schematic cartoon. (b) One-dimensional Lennard–Jones potential. (c) Two-dimensional representation with contour lines as a function of the distance x from the surface and the separation y between the two atoms. (See color insert.)

The crossing point of the two potential curves in Fig. 3.3b is now replaced by the transition state #, and its position along the reaction coordinate ϱ determines the preferred degree of freedom for overcoming the barrier. If the transition state is more to the right (early barrier), then kinetic energy of the particle incoming along the x-coordinate will be helpful, while a late barrier requires excitation along the y-coordinate of the intramolecular vibration that will be assisted by the surface temperature. The situation for dissociative adsorption of H_2 on Ni(1 1 0) and Ni(1 1 1) surfaces at zero coverages is illustrated in Fig. 3.4 [5]. In both cases, the sticking coefficient s_0 is independent of temperature. For Ni(1 1 0), it is close to unity and also not affected by the normal component of the kinetic energy of the impinging molecules, $\langle E_\perp \rangle$, while for Ni(1 1 1), it increases continuously with $\langle E_\perp \rangle$. These data suggest that in the latter case there exists a small activation

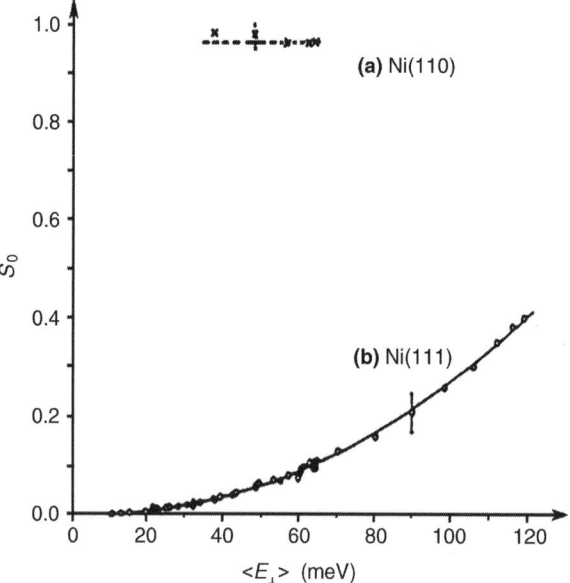

FIGURE 3.4. The sticking coefficient for dissociative adsorption of H_2 on Ni(110) and Ni(111) surfaces as a function of the kinetic energy of the impinging molecules [5].

barrier. This cannot be overcome by coupling to the heat bath of the solid since the interaction time during direct collision is too short, but requires instead high enough translational energy of the incident molecules. Clearly, for Ni(1 1 0) such a barrier is negligible as s_o is close to unity anyway. These conclusions are confirmed by theoretical calculations for this system, as reproduced in Fig. 3.5 [6]: The contour plots exhibit for Ni(1 1 1) an early barrier in the entrance channel, while there is no barrier for Ni(1 1 0).

This example clearly demonstrates the failure of TST because the time of interaction and energy exchange is not long enough. These effects were studied in detail with the dissociative

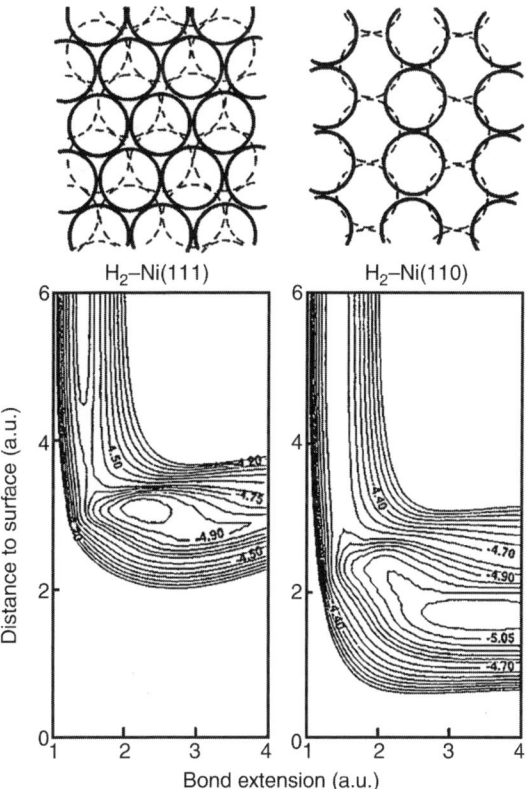

FIGURE 3.5. Theoretical contour plots for the dissociative adsorption of H_2 on Ni(1 1 1) and Ni(1 1 0) surfaces [6].

adsorption of H_2 on Cu surfaces. Molecular beam experiments were performed with systematic variation of the translational energy and combined with analysis of the populations of vibrational and rotational states, the latter even with selection of the polarization ("cartwheel" versus "helicopter") [7–9]. The detailed experimental data prompted the evaluation of sophisticated potential energy surfaces and theoretical modeling of the dynamics of this prototype reaction [10].

The dissociative adsorption of methane is another example that attracted interest [11]. Again, the dissociation probability increases with the translational energy normal to the surface, while the surface temperature is without any noticeable influence. Excitation of the bending and umbrella vibrational modes had a similar effect as increasing the translational energy. This suggests that in fact deformation of the methane molecule during collision is decisive and explains why preceding optical excitation had little effect [17]. On the other hand, if NO is excited to higher vibrational states, its dissociation probability on Cu(1 1 1) was found to be considerably higher than from the ground state [15].

Another example that is of technological relevance demonstrates the actual complexity. In catalytic ammonia synthesis, dissociative adsorption of nitrogen is the rate-limiting step where the Fe(1 1 1) surface exhibits the highest activity [14]. The sticking coefficient is very low and increases with surface temperature from which a mechanism involving a molecular precursor was concluded [15]. With molecular beam experiments, on the other hand, a marked increase of the sticking coefficient with translational energy of the impinging N_2 molecules was found, suggesting the operation of a collision-induced mechanism [16]. This puzzle could be solved by a detailed theoretical analysis [17]. In fact, there exist two channels for dissociation: (i) a precursor mediated with a low-energy barrier, but a high-entropy barrier (which explains the low sticking coefficient) dominates under usual ammonia synthesis conditions so that thermal activation takes place in the framework of TST, and (ii) a direct channel with high-activation barrier that dominates the molecular beam experiments with the high kinetic energy of the incident molecules.

Even nondissociative (molecular) adsorption may be accompanied by an activation barrier if, for example, the reaction proceeds from a physisorbed state into a chemisorbed state (trapping-mediated adsorption). In the system $O_2/Pt(1 1 1)$, for example, the O_2 molecule may be chemisorbed either in a superoxo-like or a peroxo-like state [18]. It was found that with low kinetic energies of the incident molecules, both types of surface species are formed, while at higher kinetic energies the more strongly held peroxo-like species is favored, thus reflecting correlations between incident translational energy and preferred trajectories for adsorption [19].

3.4. COLLISION-INDUCED SURFACE REACTIONS

Apart from sticking, impact of particles from the gas phase may also directly influence surface reactions. For example, not only the dissociative adsorption of methane may be promoted by the energy of the impinging CH_4 molecule but an already present adsorbed layer may also be brought to dissociation by impact of Ar atoms ("chemistry with a hammer") [20,21]. In this case, the reaction probability does not simply scale with the normal component of the kinetic energy of the Ar atoms. It was concluded that the collision energy is first transferred to the adsorbed CH_4 molecule where it is redistributed and causes dissociation or desorption.

Similar effects were found with hydrocarbons on $Au(1 1 1)$ and bombarded by Xe atoms [22], as well as with chemisorbed systems [23–25]. For O_2 chemisorbed onto $Ag(1 1 0)$ or $Pt(1 1 1)$, the impact of Xe atoms causes desorption and dissociation with equal threshold energies of 0.9 and 1.2 eV, respectively. These values are considerably higher than the respective thermal activation energies. Equal thresholds for both desorption and dissociation are presumably the consequence of rapid energy exchange between the different modes of nuclear motion (viz., the O–O vibration leading to dissociation and the M–O_2 vibration leading to desorption).

3.5. "Hot" Adparticles

The "precursor" model for adsorption on an already partly covered surface was introduced in Section 1.3. It assumes that an incident particle may also be weakly trapped on an already occupied site where it becomes thermally accommodated. There are, however, a series of observations whereafter even particles in the first layer are not immediately in thermal equilibrium but require some finite relaxation time for energy exchange with the surface during which novel dynamic effects may arise.

Even in the case of nondissociative adsorption, the tangential component of the kinetic energy of an adsorbed atom may survive considerably longer than the normal component [28]: An incident particle will first hit the repulsive part of the interaction potential and then exchange the adsorption energy only stepwise. As the variation of the potential parallel to the surface is generally weaker than perpendicular to the surface, the particle will travel some distance across the surface before coming to rest. If along this path, however, it hits an already accommodated other particle with the same mass, energy transfer will be very efficient and the two particles will preferably remain attached to each other.

Experimental verification of this effect is demonstrated by the scanning tunneling microscopy (STM) image of Fig. 3.6a [29]. About 2% of a monolayer (ML) of O_2 molecules were adsorbed on an Ag(1 1 0) surface at 65K. At this low temperature, no dissociation takes place and the mobility of accommodated particles is negligible. If the randomly impinging molecules, however, come to rest at their point of impact, most of them would show up as

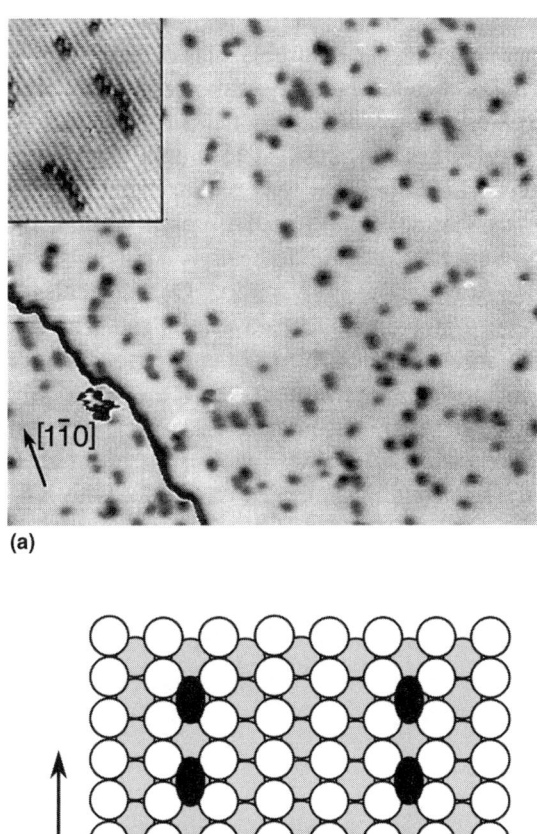

FIGURE 3.6. STM data from (nondissociative) adsorption of O_2 on a Ag(1 1 0) surface at 65K, demonstrating the operation of the hot adparticle mechanism [29]. (a) STM image after adsorption of about 2% of a monolayer. (b) Model with adsorbed O_2 molecules (black) and Ag atoms from the first (white) and second (gray) layers.

isolated species, which is actually not the case. The inset of Fig. 3.6a shows a section with atomic resolution in which both the rows of Ag atoms along the [1$\bar{1}$0] orientation as the adparticles (black dots) are discernible. Instead of isolated single adparticles, agglomerates are rather visible with typically two to four O_2 molecules with mutual separation of $2a$ along the [1$\bar{1}$0] direction, as sketched in Fig. 3.6b. This observation demonstrates the operation of the discussed "hot" adparticle effect.

Such effects may not only result from the incomplete accommodation of particles trapped on the surface from the gas phase but may also be the consequence of a reaction taking place on the surface.

The existence of "hot" adatoms, for example, is demonstrated by the STM observations on the thermally activated dissociation of O_2 chemisorbed as peroxo-type species on a Pt(1 1 1) surface leading to the formation of pairs of chemisorbed O atoms [26,27]. If the temperature is low enough ($\leq$180K), these atoms are not mobile once they are in thermal equilibrium with the solid. Interestingly, it is found that the adatoms resulting from the dissociation of an individual O_2 molecule are not located on adjacent adsorption sites but are separated from each other by 0.5–0.8 nm. This effect is a consequence of the finite time needed for damping the adsorption energy into the solid. Figure 3.3 indicates that this takes place while the two atoms are separating from each other. From the distance traveled and from the mean velocity (as determined by the adsorption energy released) in this case a mean relaxation time of about 3×10^{-13} s can be estimated during which TST will not be fulfilled.

A quite dramatic effect of this kind (which is still not yet fully understood theoretically) has been observed with the dissociative adsorption of oxygen on an Al(1 1 1) surface where the "hot" adatoms travel up to 10 nm across the surface before they come to rest [30]. An alternative explanation consists in the assumption of an abstraction process, where only one O atom remains on the surface while the other one is ejected into the gas phase. This mechanism has indeed been confirmed experimentally through detection of the O atoms released by resonance-enhanced multiphoton ionization (REMPI) [53,54]. It was found that the ejected O atoms have a mean translation energy of about 0.4 eV and it was suggested that these result from preferential dissociation from an end-on configuration of the incident molecule.

These "hot" adatoms are more energetic than they would be if accommodated with the surface, and hence they are expected to be more reactive. Experimental evidence for such effects involving adsorbed oxygen was presented by Roberts et al. [31,32], and becomes even more evident in several studies on the oxidation of CO on Pt. Figure 3.7 shows temperature-programmed reaction spectroscopy (TPRS) data for the CO_2 evolution from a Pt(1 1 1) surface with coadsorbed oxygen and carbon monoxide [33]. O_2 chemisorbed at low temperatures dissociates upon heating to 150K. When this (thermally accommodated) species was afterward exposed to CO, the TPRS data show a single peak (β) for CO_2

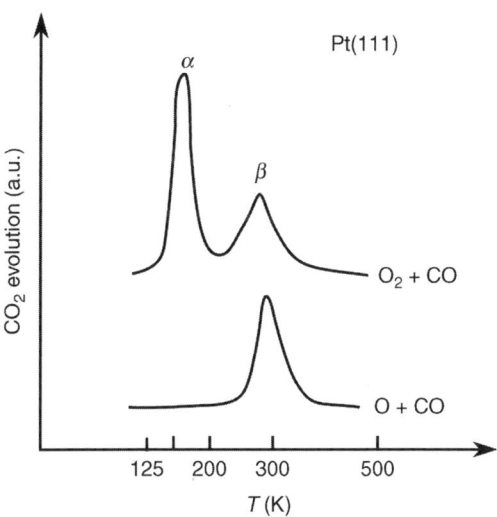

FIGURE 3.7. TPRS of CO_2 evolution from a Pt(1 1 1) surface covered by CO + O (lower trace) and CO + O_2 (upper trace) [33].

evolution at about 300K, arising from the standard reaction O_{ad} + $CO_{ad} \rightarrow CO_2$. When O_2 and CO were coadsorbed at low temperature, subsequent heating caused the appearance of another CO_2 peak (α) exactly at the temperature (150K) at which the molecularly adsorbed O_2 dissociates. There is a possibility that the "hot" adatom thermalizes before it hits a neighboring CO molecule giving rise to the "normal" β-state of CO_2 formation. However, if the "hot" O adatom hits a neighboring CO molecule while still energetically excited, immediate formation and release into the gas phase will occur (α-state) [33].

3.6. Particles Coming Off the Surface

The energy and momentum content of particles coming off the surface reflect the exchange of energy between the different degrees of freedom during interaction with the surface. The simplest case with full thermal accommodation is found in thermal desorption. Without interactions between the adsorbed particles, the rate follows first-order kinetics

$$-\frac{dn_s}{dt} = k_d \cdot n_s \qquad (3.2)$$

with the rate constant $k_d = \nu_d \cdot e^{-E_d/RT}$ being equal to the inverse of the surface residence time $\tau = 1/k_d$. Application of TST yields that $k_d = s \cdot k_{TST}$ with s being the sticking coefficient.

τ may be determined in molecular beam experiments by modulating the beam with frequency ω and recording the phase lag tan ϕ between the primary and the scattered beam: tan $\phi = \omega \cdot \tau$. Figure 3.8 represents data for the system NO/Pt (1 1 1) derived in this way [34] from which the kinetic parameters ν_d and E_d were derived.

The conclusion that chemisorbed particles are in thermal equilibrium with the solid while coupled to the surface may, however, no longer be valid if the molecules are released into

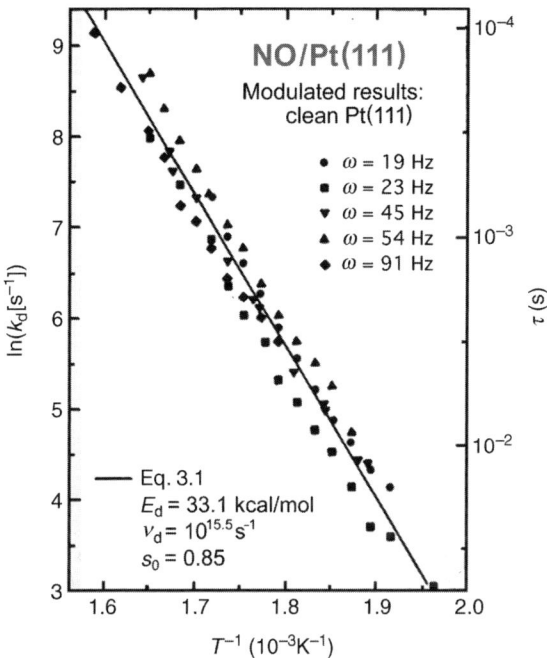

FIGURE 3.8. The mean residence time τ of NO molecules adsorbed on a Pt(1 1 1) surface as a function of temperature T [34].

the gas phase and the energy contents of the various degrees of freedom after desorption are considered. This was verified, for example, for the CO/Ru(0 0 0 1) system for which the translational energy of (thermally) desorbing molecules was evaluated from time-of-flight (TOF) measurements [35]. It turned out that the translational temperature was always considerably smaller than the surface temperature.

This effect of "cooling in desorption" is quite common [28] and extends also to other degrees of freedom. For example, Fig. 3.9 demonstrates that the mean rotational temperature T_{rot} of NO molecules desorbing from a Pt(1 1 1) surface equals the surface temperature only up to about 400K and then levels off because of incomplete excitation of the rotational motion during the desorption process [36]. As a consequence, this effect also means that rapidly rotating molecules exhibit a lower sticking coefficient.

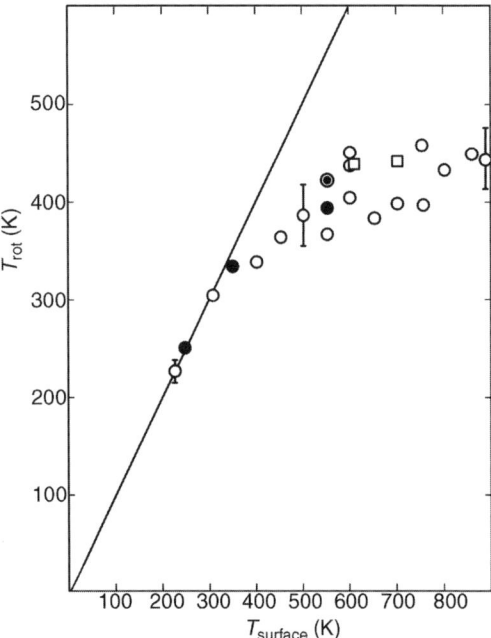

FIGURE 3.9. Variation of the rotational temperature T_{rot} of NO molecules desorbing from a Pt(1 1 1) surface as a function of surface temperature $T_{surface}$ [36].

If the adsorption proceeds across a barrier by direct collision, then the desorbing particles are expected to exhibit a mean translational energy that is higher than that corresponding to the surface temperature. In addition, the angular distribution of the desorbing particles deviates from the thermal cosine law and is peaked along the surface normal as usually described by a $\cos^n\Theta$ dependence ($n > 1$). For example, associative desorption of H_2 from a Ni(1 1 0) surface shows a $\cos\Theta$ distribution, while that from Ni(1 1 1) follows a $\cos^5\Theta$ distribution [5], just reflecting the dynamics of dissociative adsorption as described above.

More detailed insight into the energy exchange is obtained by the state-resolved molecular beam experiments [37]. As an example for these type of experiments, Fig. 3.10a shows TOF distributions from NO molecules in various rotational states (J'') coming

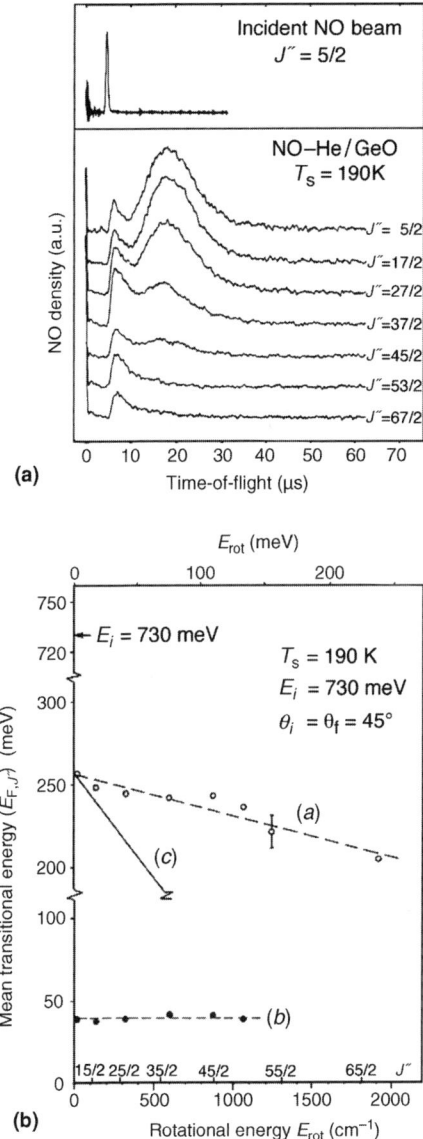

FIGURE 3.10. Scattering of NO from an oxidized Ge surface [37]. (a) Time-of-flight distributions of the incident and scattered molecules in various rotational states J''. (b) Correlation between the mean translational and the rotational energy for molecules undergoing direct inelastic scattering (a) and trapping desorption (b).

off an oxidized Ge surface after scattering of a rotationally cold molecular beam with narrow translational energy distribution centered at 730 meV [38]. The TOF data (Fig. 3.10a, lower panel) exhibit two maxima with short and long mean flight times (i.e., high and low kinetic energies) arising from molecules originating either from direct inelastic scattering or from trapping/desorption. The latter exhibits a mean translational energy of 45 meV corresponding to the surface temperature T_s and independent of the rotational states. The population of the latter follows a Boltzmann distribution with $T_{rot} = 190$K that again equals the surface temperature. The fast molecules, on the other hand, do not follow a Boltzmann distribution of their rotational populations, but rather exhibit a "rotational" rainbow with overpopulation of the higher J''. Their kinetic energy decreases linearly with the rotational energy (Fig. 3.10b, curve (*a*)), but less than when the rotational energy solely originates from the translational energy (this would give rise to curve (*c*)). A theoretical analysis shows good agreement with these experimental data [39].

Even particles being the products of exothermic reactions of thermally accommodated surface species may carry off excess energy into the gas phase.

If, for example, NO_2 decomposes at a Ge surface, the rotational distribution of the NO molecules formed is non-Boltzmann and independent of surface temperature, suggesting that the NO molecule is directly ejected into the gas phase upon abstraction of the second O atom by the surface [40].

Evidence for the formation of hyperthermal molecules coming off the surface is obtained by recording their angular (and velocity) distributions [41], as exemplified by the catalytic oxidation of CO on a Pt(1 1 1) surface, a reaction that will be discussed in detail in Section 6.3. The product molecule CO_2 is only weakly held to the surface and thus experiences a repulsive potential after passing the transition state. As a result, the angular distribution is strongly peaked, and the kinetic energy can increase to

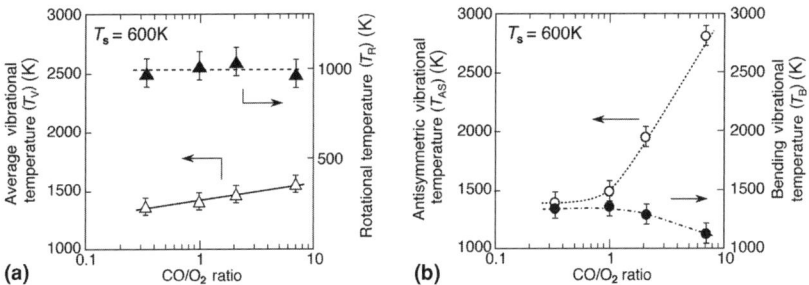

FIGURE 3.11. Chemiluminescence in the CO oxidation on a Pd(1 1 0) surface [46]. (a) Average vibrational and rotational temperatures of CO_2 as a function of the CO/O_2 ratio at $T_s = 600K$. (b) Temperatures of the antisymmetric and the bending vibrations of CO_2 as a function of the CO/O_2 ratio.

330 meV—about the 10-fold thermal value given by the surface temperature [42]. The transition state is presumably bent while the product molecule is linear, and hence noticeable excitation of internal degrees of freedom is observed by IR chemiluminescence [43–45]. As an example, Fig. 3.11 shows for a Pd(1 1 0) surface the average vibrational and the rotational temperaturesas well as the temperatures of the asymmetric and the bending vibrations as a function of the CO/O_2 ratio at a surface temperature of 600K [46]. These data indicate clearly that the asymmetric stretch vibration is considerably more excited than the other vibrational modes.

3.7. ENERGY EXCHANGE BETWEEN ADSORBATE AND SURFACE

Thermal equilibrium between adsorbate and surface means that both subsystems exhibit the same temperature. Infrared spectra from CO adsorbed on a Ru(0 0 0 1) surface represent a convenient example for monitoring the temperature of the adsorbate. Figure 3.12 shows a series of IR spectra from CO adsorbed on a Ru(0 0 0 1) surface at different temperatures [47]. With increasing T, there is a continuous shift of the band toward lower wave numbers together with a continuous line broadening. This effect

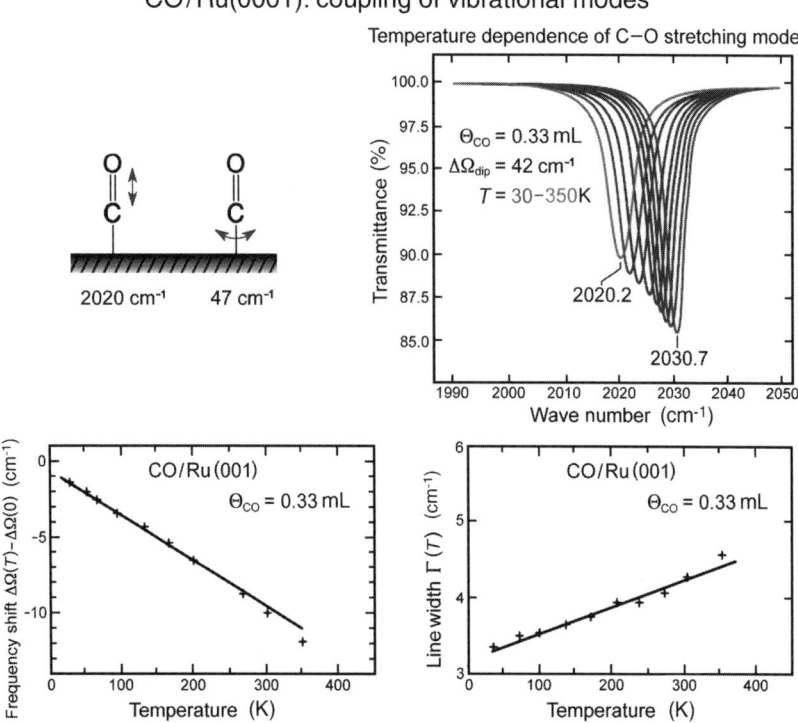

FIGURE 3.12. Variation of the IR spectra from CO adsorbed at a Ru(0001) surface with surface temperature due to coupling of the stretch mode to that of frustrated rotation [47].

has to be attributed to coupling of the C–O stretch vibration to the low-energy frustrated rotation that becomes thermally more excited at higher temperature. In this way, the C–O stretch vibration offers a convenient means to monitor the temperature of the adsorbate. To record the temporal variation of the adsorbate temperature upon rapid heating, this band was recorded by means of time-resolved broadband sum frequency spectroscopy (SFG) [48], which in our experiment reaches a time resolution of 0.5 ps [49]. Heating of the surface was achieved by absorption of a rapid (110 fs) IR laser pulse by the conduction electrons of the metal that couple to the lattice vibrations (phonons) and to the adsorbate. By varying the delay time

Energy Exchange Between Adsorbate and Surface 71

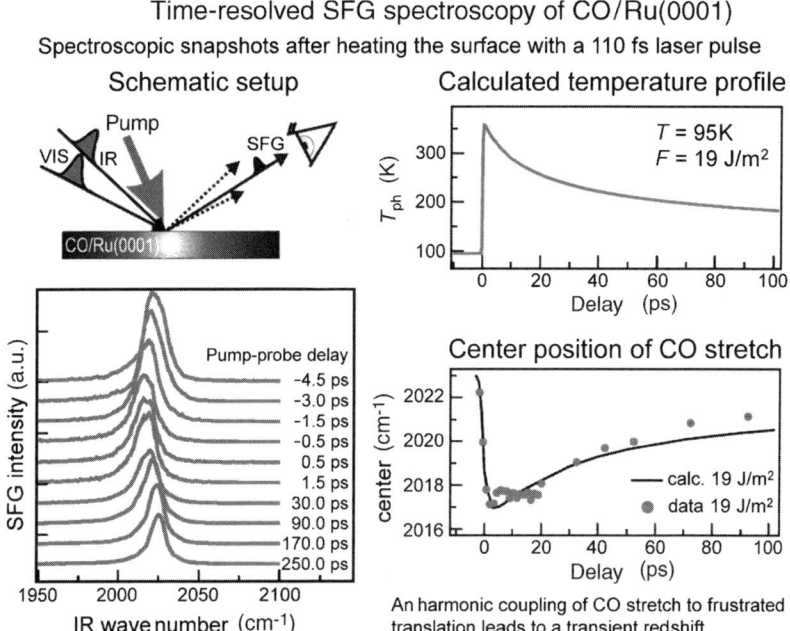

FIGURE 3.13. Time-resolved SFG spectroscopy from CO adsorbed on a Ru (0 0 0 1) surface after rapid heating by an intense IR pulse and subsequent cooling [49].

between pump and probe pulse in the SFG experiments, a series of spectra at varying delay times after heating could be recorded. The results are reproduced in Fig. 3.13 when the surface was heated from 95 to 350K. The CO band experiences a continuous redshift due to heating, which returns upon cooling. The variation of the band position with time agrees well with that expected if the adsorbate temperature would follow that calculated from the thermal parameters of the surface. This result demonstrates that the adsorbate temperature follows that of the substrate on the timescale of about 1 ps. If the same experiment is extended up to 800K, partial desorption takes place and the CO band disappears almost completely due to coupling of the molecule tumbling around on the surface to the other thermally excited degrees of freedom.

More information about typical time constants for the exchange of energy between the degrees of freedom of the adsorbate and the solid substrate could again be obtained by rapidly heating up the conduction electrons of the metal by a fs IR pulse [50]. A Ru (0 0 0 1) surface was covered by dissociatively adsorbed H (D) atoms that could subsequently be associatively desorbed by thermal desorption spectroscopy with a mean (thermal) kinetic energy of about 350K. If, on the other hand, the surface was irradiated with 130 fs long IR pulses at 800 nm, the desorbing molecules exhibited much higher kinetic energies around 2000K, and the yield of H_2 with a single shot was about 10-fold that for D_2. Further experiments and detailed analysis provided the following picture, as illustrated in Fig. 3.14 [51]: Absorption of the photons by the conduction electrons creates hot electrons above the Fermi

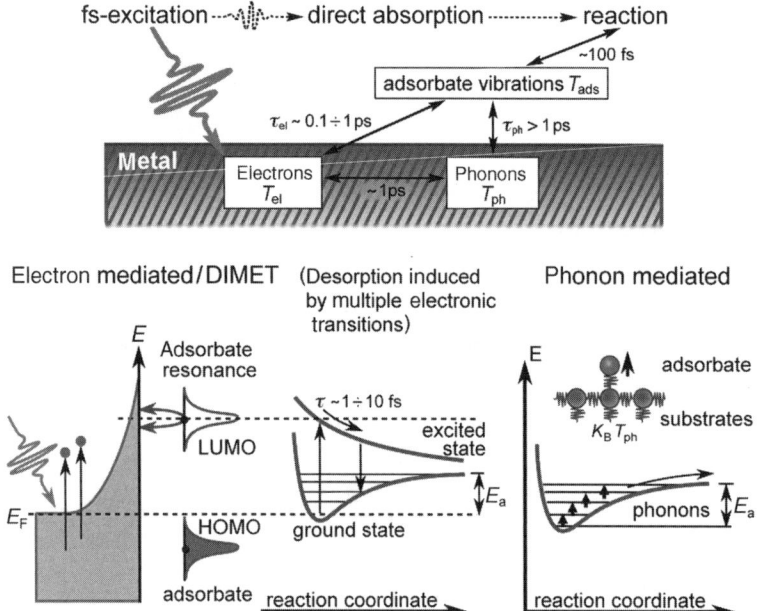

FIGURE 3.14. The mechanism of desorption caused by hot electrons (DIMET) created by absorption of a fs IR pulse versus phonon-mediated desorption from the ground state [51]. (See color insert.)

level E_F that rapidly (<100 fs) equilibrate internally to an electron temperature T_{el} and then couple to the lattice with its phonon temperature T_{ph}. In contrast to phonon-mediated desorption from the ground state (where the system climbs up the phonon levels in the ground-state potential), an adsorbate-derived affinity level above E_F may transiently be populated, where the system is transferred into an electronically excited state. The gradient on this potential causes nuclear motion, leading eventually by multiple excitation–relaxation steps to desorption via a nonadiabatic mechanism (DIMET) based on coupling between electronic and nuclear degrees of freedom. Because of the mass difference between H and D, the motion of the latter species on the excited potential is slower and hence the pronounced isotopic effect in the desorption yield. Figure 3.15 depicts the variation of the different temperatures (T_{el}, T_{ph}, and T_{ad}) with time for two subsequent pulses. They all equilibrate within about 1 ps to the common lattice temperature T_{ph} signaling thermal equilibrium.

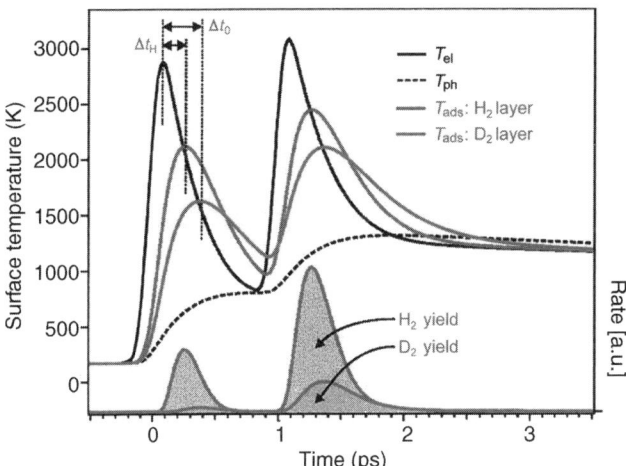

FIGURE 3.15. Variation of the electronic, phonon, and adsorbate temperatures (T_{el}, T_{ph}, and T_{ads}) with time following irradiation of H (D) covered Ru(0 0 0 1) surface by two subsequent fs IR pulses [51]. (See color insert.)

An example for which separation of the timescales for electron and phonon excitations directly influenced the catalytic reaction was found with CO oxidation on a Ru(0001) surface onto which O + CO had been coadsorbed [52]. Upon increasing the temperature of the sample, only desorption of CO but no formation of CO_2 took place. However, when the rapid IR pulses (130 fs at 800 nm) were applied, the situation changed: Now both CO and CO_2 were released into the gas phase. Two pulse correlation experiments (similar as with Fig. 3.15) with varying delay times between two subsequent pulses revealed that the relaxation time for the decay of the excitation responsible for CO desorption was of the order of about 20 ps, but was much shorter (~1 ps) for CO_2 formation. As illustrated in Fig. 3.16, CO_2 formation essentially takes place during the first picosecond

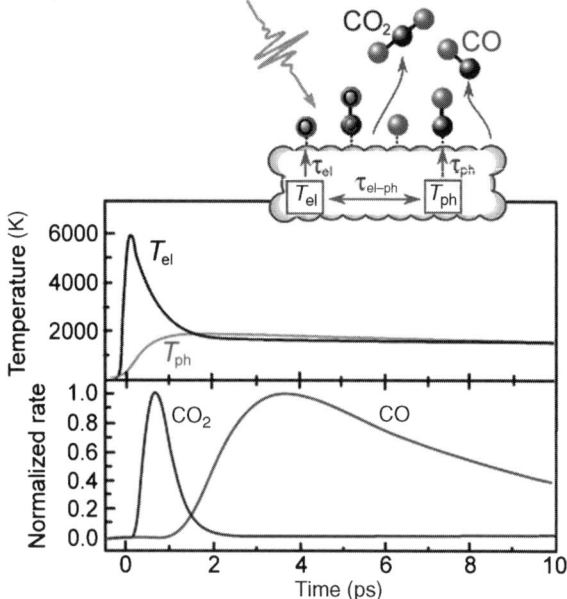

FIGURE 3.16. Temporal evolution of a Ru(0001) surface covered by coadsorbed O + CO after irradiation with an IR pulse of 130 fs duration. Variation of the electronic and phonon temperatures, T_{el} and T_{ph}, respectively, with time leading to CO desorption and CO_2 evolution [52].

after absorption of the IR by the conduction electrons when T_{el} reaches values up to 6000K, while desorption of CO starts later only with the increase of T_{ph}. The latter process occurs with a (thermal) activation energy that is considerably smaller than that for CO_2 formation. This explains why with normal heating CO desorption takes place before the recombination of CO + O. The formation of hot electrons above E_F during the laser shot, however, causes appreciable population of an O 2p-derived level that is antibonding with respect to the adsorbate–substrate bond, where coupling to nuclear motion then initiates CO_2 formation. For adsorbed CO, on the other hand, the lowest lying empty level (derived from the CO-$2\pi^*$ orbital) is located about 5 eV above E_F, and is thus much too high in energy to become substantially populated by hot electrons. Hence, CO desorption takes place from the electronic ground state through coupling to phonon excitations.

References

1. H. Eyring and M. Polanyi, *Z. Phys. Chem. B* **12** (1931) 279.
2. T. Engel, *J. Chem. Phys.* **69** (1978) 373.
3. J. P. Toennies, *Surface Phonons*, Springer Series in Surface Science, Vol. 21, Springer, 1991, Chapter 5.
4. G. Comsa, *Surf. Sci.* **299/300** (1994) 77.
5. H. Robota, W. Vielhaber, M. C. Lin, J. Segner, and G. Ertl, *Surf. Sci.* **155** (1985) 101.
6. J. K. Nørskov, P. Stoltze, and U. Nielsen, *Catal. Lett.* **9** (1991) 173.
7. C. T. Rettner, H. A. Michelsen, and D. J. Auerbach, *J. Chem. Phys.* **102** (1995) 4625.
8. K. D. Rendulic and A. Winkler, *Surf. Sci.* **299/300** (1994) 261.
9. D. Wetzig, M. Rutkowski, R. David, and H. Zacharias, *Europhys. Lett.* **36** (1996) 31.
10. A. Groß, *Surf. Sci. Rep.* **32** (1998) 291.
11. S. T. Ceyer, *Annu. Rev. Phys. Chem.* **39** (1988) 479.

12. J. T. Yates, J. J. Zinck, S. Sheard, and W. H. Weinberg, *J. Chem. Phys.* **70** (1979) 2266.
13. H. Hou, S. J. Coulding, C. T. Rettner, A. M. Wodtke, and D. J. Auerbach, *Science* **277** (1997) 80.
14. G. Ertl, in: *Encyclopedia of Catalysis* (ed. I. T. Horvath), Vol. **1**, Wiley, 2003, p. 329.
15. G. Ertl, S. B. Lee, and M. Weiss, *Surf. Sci.* **114** (1982) 515.
16. C. T. Rettner and H. Stein, *Phys. Rev. Lett.* **59** (1987) 2768.
17. J. J. Mortensen, L. E. Hansen, B. Hammer, and J. K. Nørskov, *J. Catal.* **182** (1999) 479.
18. C. Paglia, A. Nilsson, B. Hernass, O. Karris, P. Bennik, and N. Martensson, *Surf. Sci.* **342** (1995) 119.
19. P. D. Nolan, B. R. Lutz, P. L. Tanaka, J. E. Davis, and C. B. Mullins, *Phys. Rev. Lett.* **81** (1998) 3179.
20. J. D. Beckerle, Q. Y. Yang, A. D. Johnson, and S. T. Ceyer, *J. Chem. Phys.* **86** (1987) 7236.
21. J. D. Beckerle, A. D. Johnson, Q. Y. Yang, and S. T. Ceyer, *J. Chem. Phys.* **91** (1989) 5756.
22. J. Libuda and G. Scoles, *J. Chem. Phys.* **192** (2000) 1522.
23. G. Szulczewski and R. J. Lewis, *J. Chem. Phys.* **98** (1993) 5974; **101** (1994) 104.
24. C. Akarlund, I. Zoric, and B. Kasemo, *J. Chem. Phys.* **104** (1994) 7359; **109** (1998) 866.
25. C. Akarlund, I. Zoric, B. Kasemo, A. Culpolillo, F. Buatier de Mongeot, and M. Rocca, *Chem. Phys. Lett.* **270** (1997) 157.
26. J. Wintterlin, R. Schuster, and G. Ertl, *Phys. Rev. Lett.* **72** (1996) 123.
27. B. C. Stipe, M. A. Rezai, and W. Ho, *J. Chem. Phys.* **197** (1997) 6443.
28. C. T. Tully, *Surf. Sci.* **111** (1981) 461.
29. J. V. Barth, T. Zambelli, J. Wintterlin, and G. Ertl, *Chem. Phys. Lett.* **270** (1997) 152.
30. H. Brune, J. Wintterlin, R. J. Behm, and G. Ertl, *Phys. Rev. Lett.* **68** (1992) 624.
31. C. T. Au and M. W. Roberts, *Nature* **319** (1986) 206.
32. A. F. Carley, P. R. Davies, and M. W. Roberts, *Phil. Trans. Roy. Soc. A* **363** (2005) 820.
33. T. Matsushima, *Surf. Sci.* **127** (1983) 403.
34. C. T. Campbell, G. Ertl, and J. Segner, *Surf. Sci.* **115** (1982) 309.

35. S. Funk, M. Bonn, D. Denzler, C. Hess, M. Wolf, and G. Ertl, *J. Chem. Phys.* **112** (2000) 9888.
36. J. Segner, H. Robota, W. Vielhaber, G. Ertl, F. Frenkel, J. Häger, W. Krieger, and H. Walther, *Surf. Sci.* **131** (1983) 273.
37. J. A. Barker and D. J. Auerbach, *Surf. Sci. Rep.* **4** (1985) 1.
38. A. Mödl, T. Gritsch, F. Budde, T. J. Chuang, and G. Ertl, *Phys. Rev. Lett.* **57** (1986) 384.
39. C. W. Muhlhausen, L. R. Williams, and J. C. Tully, *J. Chem. Phys.* **83** (1985) 2594.
40. A. Mödl, H. Robota, J. Segner, W. Vielhaber, M. C. Lin, and G. Ertl, *Surf. Sci.* **169** (1986) L341.
41. T. Matsushima, I. Rzeznicka, and Y. Ma, *Chem. Rec.* **5** (2005) 81.
42. K. H. Allers, H. Pfnür, P. Feulner, and D. Menzel, *J. Chem. Phys.* **100** (1994) 3985.
43. D. A. Mantell, S. B. Ryali, B. L. Halpern, G. L. Haller, and J. B. Fenn, *Chem. Phys. Lett.* **81** (1981) 185.
44. C. Wei and G. L. Haller, *J. Chem. Phys.* **105** (1996) 810.
45. H. Uetsuka, K. Watanabe, and K. Kunimori, *Surf. Sci.* **363** (1996) 73.
46. K. Nabao, S. Ito, K. Tamishige, and K. Kunimori, *Chem. Phys. Lett.* **410** (2005) 86.
47. P. Jakob and B. N. J. Persson, *Phys. Rev. B* **56** (1997) 10644.
48. L. J. Richter, et al., *Opt. Lett.* **23** (1998) 1594.
49. M. Bonn, C. Hess, S. Funk, J. H. Miners, B. N. J. Persson, M. Wolf, and G. Ertl, *Phys. Rev. Lett.* **84** (2000) 4653.
50. D. N. Denzler, C. Frischkorn, C. Hess, M. Wolf, and G. Ertl, *Phys. Rev. Lett.* **91** (2003) 226102.
51. D. N. Denzler, C. Frischkorn, M. Wolf, and G. Ertl, *J. Phys. Chem. B* **108** (2004) 14503.
52. M. Bonn, S. Funk, C. Hess, D. Denzler, C. Stampfl, M. Scheffler, M. Wolf, and G. Ertl, *Science* **285** (1999) 1042.
53. A. J. Komrowski, J. Z. Sexton, A. C. Kummel, M. Binetti, O. Weisse, and E. Hasselbrink, *Phys. Rev. Lett.* **87** (2001) 246103.
54. M. Binetti and E. Hasselbrink, *J. Phys. Chem. B* **108** (2004) 14677.

CHAPTER 4

ELECTRONIC EXCITATIONS AND SURFACE CHEMISTRY

4.1. INTRODUCTION

The primary concept of reaction dynamics is based on the Born–Oppenheimer approximation assuming that electronic motion is much faster than nuclear motion, so the electrons adjust instantaneously to the current nuclear configuration and the system evolves on a Born–Oppenheimer potential energy surface (PES). Nonadiabatic coupling effects between nuclear motions and electronic excitations are then usually neglected, although at metal surfaces a whole manifold of electron–hole pair excitations exists, and thus such nonadiabatic coupling may become more relevant [1]. The last examples of Chapter 3 were already concerned with reactions that were affected by rapid heating of the conduction electrons to states above the Fermi level by absorption of short light pulses. These effects are denoted as surface femtochemistry and are of considerable current interest [2]. In brief, absorption of a fs light pulse creates a nonequilibrium distribution of hot electrons that may be characterized by an electronic temperature T_{el} much higher than the phonon and adsorbate temperatures, T_{ph}

Reactions at Solid Surfaces. By Gerhard Ertl
Copyright © 2009 John Wiley & Sons, Inc.

and T_{ad}, respectively, eventually then initiating surface reactions. Because of the extremely short-lived character of the electronic excitations, the resulting reaction dynamics is nevertheless essentially governed by the ground-state potential energy surface.

This chapter is essentially concerned with effects characteristic of electronically excited states: (i) surface reactions causing electronic excitations, which may be directly analyzed, and (ii) controlled electronic excitations leading to surface reactions.

In molecular reactions, electronic excitations as a consequence of the chemical transformation are often associated with light emission called chemiluminescence. As outlined in Section 3.1, with metal surfaces the relaxation times of electronic excitations (from delocalized band states) are much shorter ($\sim 10^{-15}$ s) than those for photons ($\sim 10^{-8}$ s), so this decay channel will be operating only with very low probability, except in systems where the electrons are localized and their excitations exhibit long lifetimes.

Such an example was found with the agglomeration of small metal (Ag and Cu) clusters in noble gas matrices that may be accompanied by the emission of visible light [3,4]. As illustrated in Fig. 4.1, metal atoms embedded into a noble gas matrix migrate upon heating and form small clusters that eventually agglomerate into larger ones. The gain in binding energy at this step may lead to the ejection of electronically excited fragments that then decay by light emission. Figure 4.2 shows the spectra of the emitted radiation for Cu in Ne and Ar matrices, respectively, which are identified with emission from excited Cu atoms and dimers. It is obvious that for this process the lifetime of the electronically excited M^*_{m+n} intermediate has to be long enough to enable dissociation into M^* or M^*_2 fragments. It was concluded that the size of Ag clusters whose agglomeration may give rise to the observed chemiluminescence effects has to be below 10–20 atoms, above which size the picture of localized molecular orbitals is replaced by a delocalized electron gas with lifetimes of the excited states in the fs range.

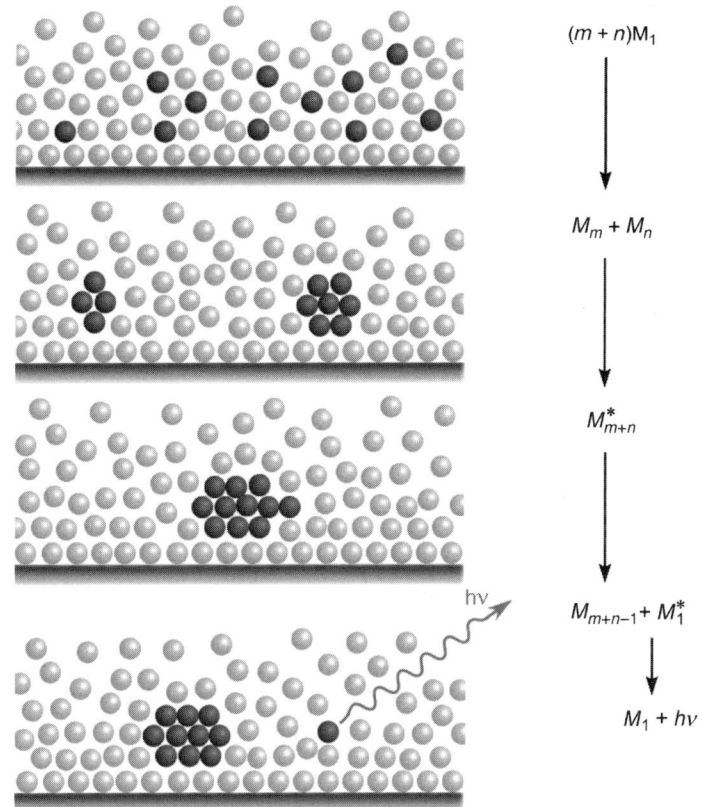

FIGURE 4.1. Agglomeration of Cu or Ag atoms in a noble gas matrix leads to electronically excited clusters that eject fluorescing atoms or dimers.

4.2. EXOELECTRON EMISSION

Langmuir and Kingdon [5] observed that thermal desorption of cesium from tungsten occurs as Cs^+ ions and concluded: "Hence cesium atoms leaving a tungsten surface are robbed of their valence electrons by tungsten." This effect forms the basis of thermoionic energy conversion and can readily be rationalized by the fact that it costs less energy to ionize a Cs atom (3.9 eV) than one gains by transferring this electron to W with its work function of over 5 eV. This example demonstrates the possibility of electron transfer in a nonadiabatic surface reaction. Less obvious is the

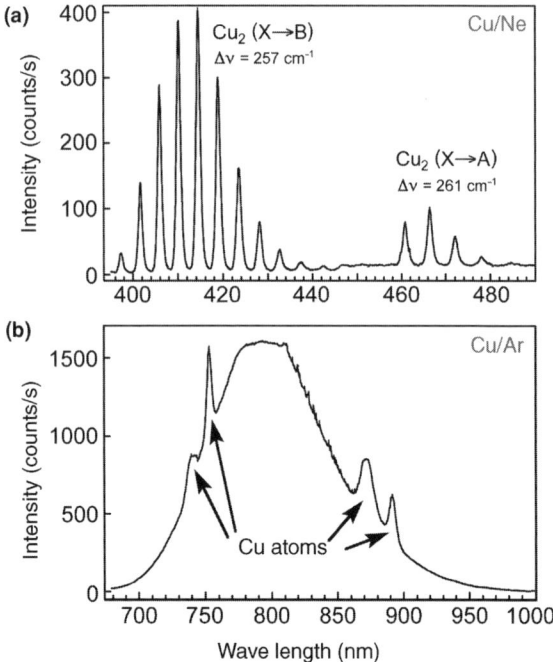

FIGURE 4.2. The spectra of light emitted from Cu_2 in a Ne matrix (a) and of Cu atoms in an Ar matrix (b) [3].

interpretation of earlier reports by Haber and Just [6], whereafter reaction of oxygen (and other electronegative molecules) with alkali metal surfaces causes the emission of electrons, whose origin obviously has to be associated with the energy gain involved in the oxidation process. This effect has been denoted "exoelectron emission" [7] and clearly demonstrates the possibility for electronic excitations by surface reactions from the electronic ground state [8].

As an example, Fig. 4.3 shows the intensity of electrons (a) and the variation of the work function (b) as a function of O_2 exposure for a Li surface reacting with oxygen [9]. The electron yield increases continuously up to a maximum and then drops sharply, and the work function ϕ decreases simultaneously continuously with progressing oxidation. The energy distribution of the emitted

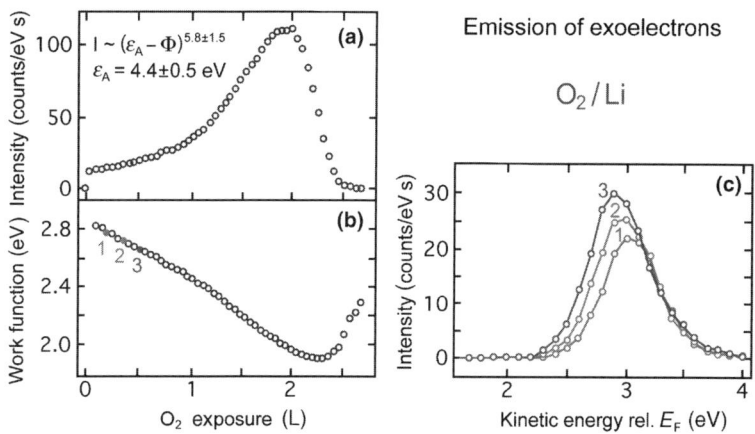

FIGURE 4.3. Emission of exoelectrons upon interaction of O_2 with a Li surface [9]. (a) Electron intensity as a function of O_2 exposure. (b) Variation of the work function with O_2 exposure. (c) Kinetic energy distributions of the emitted electrons at three stages marked in (b).

electrons with respect to the Fermi level E_F at three stages marked in Fig. 4.3b is reproduced in Fig. 4.3c. These data show that the low-energy cutoff is determined by the respective work function, thus explaining the increasing yield with decreasing work function.

This process may be rationalized by a theoretical model proposed by Kasemo et al. [10] and illustrated in Fig. 4.4: Upon approach to the surface, the lowest empty electronic state (affinity level) of the impinging particle at E_A (which for O_2 is 0.4 eV below the vacuum level E_V) is continuously lowered. When it crosses the Fermi level, there is a high probability that it becomes occupied by an electron tunneling from E_F ("harpooning"), and then the negative ion is further accelerated to the surface where bond formation takes place. There is, however, a small probability that the neutral particle reaches the surface before ionization. The empty level will then be at ε_A that will be occupied by an electron from the metal, and the energy released excites another electron via the Auger effect. This exoelectron obviously attains its maximum kinetic energy if both electrons originate from the Fermi level, namely, $E_{kin,max} = -\varepsilon_A$, while the minimum kinetic energy

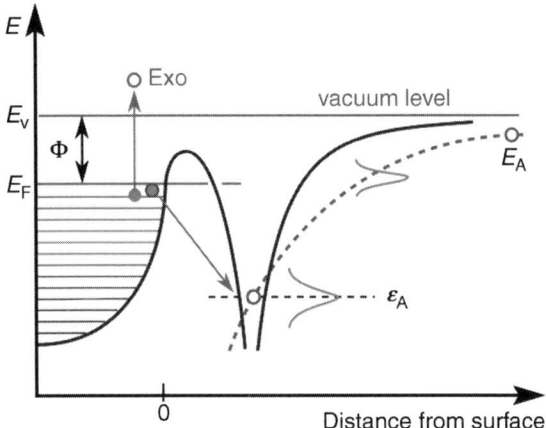

FIGURE 4.4. Potential diagram illustrating the mechanism of exoelectron emission [10].

is given by the work function, $E_{kin,min} = \phi$. This model explains why the energy distributions of Fig. 4.3c have a common high-energy edge (i.e., ε_A) while the low-energy cutoff is equal to the (varying) work function.

Such a nonadiabatic reaction pathway appears to be rather improbable since the quenching of electronic excitations at metal surfaces is usually much faster than the timescale for nuclear motion, and with the O_2/Li system, the probability for exoelectron emission is indeed $<10^{-6}$ e/incident O_2 molecule. The competition between nuclear and electronic motion is nicely reflected by the exponential increase of the electron yield with the velocity of the impinging molecules as shown in Fig. 4.5 for the system $O_2 + Cs$ [11]. Note that a velocity of 2×10^3 m/s is equivalent to a distance of 0.2 nm in 100 fs, just in agreement with the timescales for electronic relaxation.

More detailed discussion of the $O_2 + Li$ system [12] revealed that the whole reaction is indeed more complex: Electron transfer from the substrate to the incoming O_2 molecule first causes the formation of O_2^{2-} species that readily dissociate. The O^- species formed can pick up a second electron, and this latter step may be

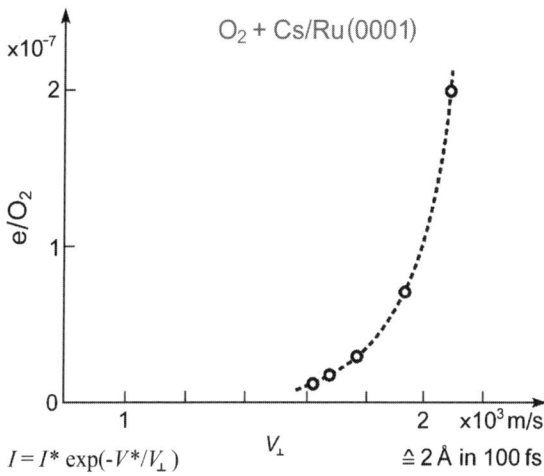

FIGURE 4.5. Variation of the initial yield of exoelectrons in the reaction of O_2 with a thin Cs film with the velocity of the incident O_2 molecules [11].

associated with exoelectron formation, as illustrated in Fig. 4.4. The intermediate formation of O^- in front of the surface is supported by the detection of ejected O^- ions (however, with very small probability, $\leq 10^{-8}\, O^-$ per incident O_2) in the oxidation of cesium. The electron affinity of O (1.46 eV) is smaller than the work function of the Cs surface (2.0 eV). The energy necessary for this electronic excitation has to originate from the exothermicity of the reaction [13]. The emitted O^- ions have practically zero kinetic energy, and a mechanism was proposed whereafter their release is a consequence of strong repulsion in the O_2^{2-} species intermediately formed in front of the surface.

Similar effects had been reported earlier for Cl^- and Br^- ions ejected from alkali metal surfaces exposed to halogens, however with considerably higher yields [14,15]. Interestingly, there seems to exist a direct relation between the yield of negative ions and the energy difference between the work function and the affinity level, $\Delta E = \phi - E_A$.

The energy scheme of Fig. 4.4 suggests that apart from Auger decay deexcitation might also occur via light emission.

Fluorescence was indeed observed for Cl_2 interacting with K surfaces, but with much lower yield than exoelectron emission, while in the reaction with O_2 the light intensity was below the detection limit [16]. This is in agreement with general experience whereafter at metal surfaces, fluorescence is strongly suppressed by Auger deexcitation for energies up to 500 eV [17].

Instead of increasing the translational energy of the impinging particle, the emission of electrons may also be enhanced by collision with highly vibrationally excited molecules as verified with NO molecules colliding with a Cs-covered Au(1 1 1) surface [74]. Interestingly, with this system, the electron yield increases with *decreasing* velocity up to 0.1 for $v = 18$, for which effect a vibrational autodetachment mechanism was proposed [75].

4.3. Internal Electron Excitation: "Chemicurrents"

Direct manifestation of electronic excitations caused by surface reactions through emission of exoelectrons will be restricted to interactions between electronegative molecules, such as O_2, NO_2, or halogens, with low work function surfaces, such as alkali metals, since the energy of the affinity-derived level ε_A has to exceed the work function (cf. Fig. 4.4). Even then, only a small fraction of excited electrons will escape since many combinations of lower lying electrons will cause excitations only to energies below the vacuum level. It is hence to be expected that nonadiabatic channels are existing for a large variety of surface reactions [18]. The occurrence of such processes of internal electron excitation by surface reaction was indeed recently demonstrated by Nienhaus et al. [19–21]. In the first experiments [19], a 7.5 nm thick Ag film had been deposited onto a Si substrate. At the buried interface, a Schottky barrier is formed by space–charge effects, as sketched in the inset of Fig. 4.6. The system now acts as a diode in which electrons with excitation energies >0.8 eV can pass from

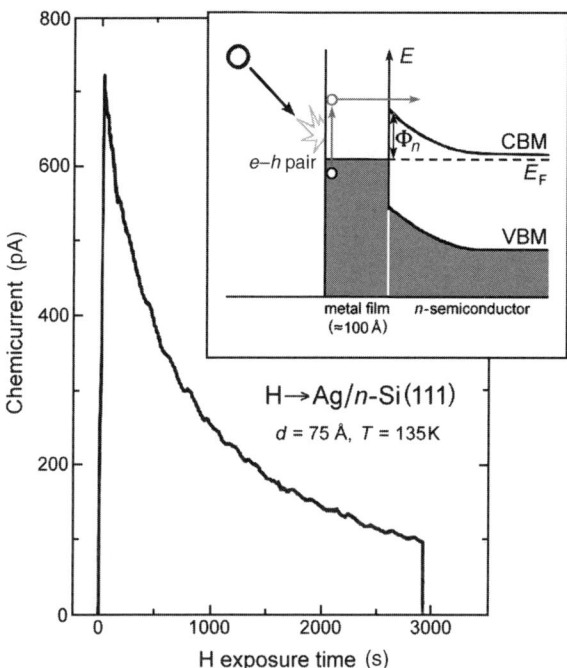

FIGURE 4.6. Mechanism for the creation of a chemicurrent in a Schottky diode formed by a thin metal film on a semiconductor substrate [19].

the metal film into the semiconductor substrate where they can be detected as macroscopic current. The data of Fig. 4.6 refer to a flux of H atoms causing initially a current of about 10^{-10} A. This effect has been denoted "chemicurrent," and an efficiency of the order 10^{-3} e/adsorption event was estimated [19]. Theoretical treatments [22–24] revealed qualitative agreement with this number as well as with the isotope effect between H and D [25].

In later experiments on the oxidation of Mg, it was demonstrated that the chemicurrent transients directly represent the reaction rate at the surface [26].

As an alternative to the diode device, Hasselbrink et al. [27] proposed a metal–insulator–metal (MIM) layer system. Electrons excited at the outer metal surface may pass through the conduction band of an insulating oxide layer to a second metal electrode.

In this way a steady-state current was recorded upon exposing a Au film to a flux of H atoms that continuously recombine to H_2 at the surface. A probability of 2×10^{-5} e/H atom was determined that is about two orders of magnitude smaller than that reported in Ref. [19], which has to be attributed to the higher energy barrier of the MIM device.

The original diode device has even been successfully applied to monitor chemicurrents under the steady-state catalytic reaction of CO oxidation [28–34]. The initially reported electron yields of three electrons for the production of four CO_2 molecules [29,30] were unphysically high. In the meantime, yields of the order 10^{-3} to 10^{-4} [33,34] were determined, which values fit into the general picture. Since supported catalysts often consist of metal nanoparticles on an oxide support, it is argued that these chemicurrents may be used to monitor the electronic excitations associated with "real" catalysts.

4.4. Electron-Stimulated Desorption

Irradiation of an adsorbate-covered surface by low-energy electrons may cause desorption of positive as well as negative ions from the adsorbate [35]. This effect is denoted as desorption induced by electronic transitions (DIET) and is usually discussed [36] in terms of the so-called MGR model [37,38]. The model assumes that the incident electron initiates a Franck–Condon transition of the adsorbate–surface system from the ground state to an excited state from where (during its short lifetime) the adsorbate or one of its fragments may escape from the surface. This model was later somewhat modified by Antoniewicz [39] who assumed that the adsorbate, by interaction with the incident electron, becomes instantaneously positively ionized. This ion then experiences a screened image potential that attracts it toward the surface. The closer distance to the surface enhances the probability for reneutralization, the system returns to the ground-state

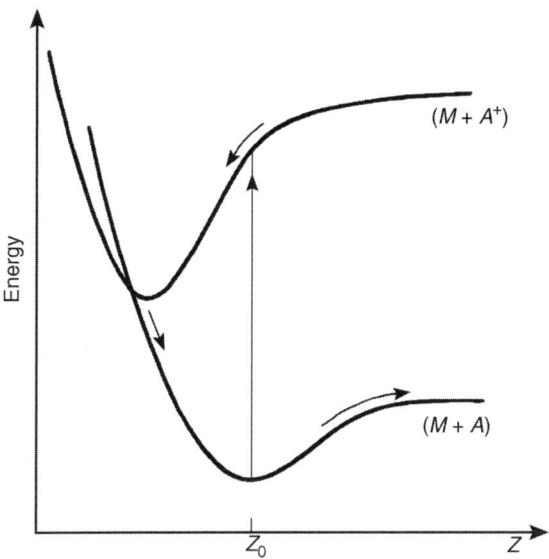

FIGURE 4.7. The Antoniewicz model for electron-stimulated desorption (ESD) of neutral particles [39].

potential from where the adsorbate is repelled and eventually desorbed. This simple situation for neutral desorption is illustrated in Fig. 4.7, while ionic desorption requires a somewhat more complex model.

The experiments may involve measurements of the desorption yield, energy dependence, and identification of the desorbing species. As an example, the effects found with PF_3 adsorbed on Pt will be briefly discussed [40,41]. Electron bombardment of an adsorbed monolayer mainly gives rise to an F^+ signal with a threshold energy for the electrons of 26.5 eV while the F^+ ions exhibit a peak energy of 4 eV, which effect is attributed to excitation of the F 2s level. With thick adlayers, P^+, PF^+, and PF_2^+ signals were observed [40]. In addition, the ejection of negative ions, mainly F^-, was observed for which more complex mechanisms were discussed [41].

An interesting extension of the ESD technique was achieved by measuring the angular distribution of desorbing particles

(ESDIAD) [42]. This method was considerably improved by incorporation of a digital acquisition system and background subtraction [43] and provides some information about the symmetry of the adsorption site and the direction of the adsorbate bond [35].

ESDIAD from PF_3 adsorbed on a Ni(1 1 1) surface at 85K exhibited a sixfold symmetric F^+ pattern, as sketched in Fig. 4.8a [44]. Molecular adsorption at atop sites through the lone pair of P was concluded with PF bonds directed toward the nearest-neighbor Ni atoms in agreement with previous TPD and UPS experiments [45]. Prolonged exposure to the electron beam changed the F^+ ESDIAD pattern to that reproduced in Fig. 4.8b that was attributed to transformation into a mixed $PF_2 + PF$ overlayer.

Instead of an electron gun, the tip of a scanning tunneling microscope may be used as a source for electronic excitations, where the adsorbate dynamics may be probed even on atomic scale. The possibility of the STM tip to manipulate atoms or molecules on a surface had been demonstrated earlier [46,47]. Stipe et al. [48] placed an STM tip over an O_2 molecule adsorbed on a Pt(1 1 1) surface, dissociated the molecule by electrons tunneling from the tip, and subsequently imaged the two O atoms formed. After dissociation, these atoms were found to be up to three lattice constants apart from each other, just as in the case of thermal dissociation, as discussed in section 3.5 in connection with the effect of "hot" adatoms. In the next step, even the rotational motion of a single O_2 molecule was induced by the STM tip [49]. It was found that a current pulse through an O_2 molecule caused its rotation by 60° into its second equivalent orientation and so on. Above a certain bias voltage, the rate of rotation varied linearly with the tunneling current, indicating a single-electron mechanism. The energy barrier for rotation was estimated to be 0.15 eV, while that for dissociation was about 0.4 eV. These results indicate that the potential energy surface involves multidimensional

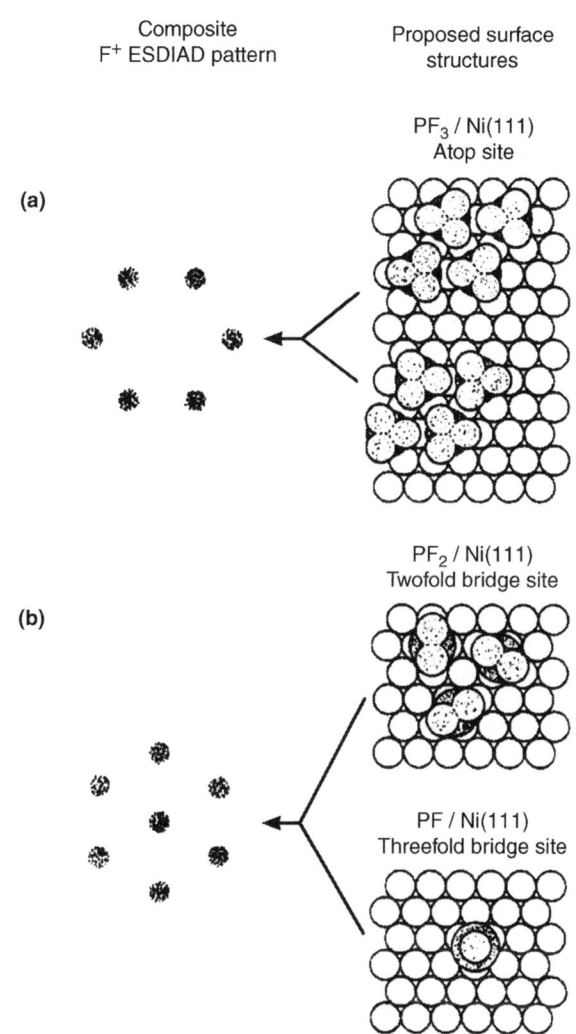

FIGURE 4.8. Electron-stimulated desorption angular distribution (ESDIAD) data from PF_3 adsorbed on a Ni(1 1 1) surface [44]. (a) Unperturbed PF_3 adlayer. (b) After electron induced fragmentation into PF_2 and PF species.

pathways with possible couplings between different modes, demonstrating how complex the full dynamics even for such an apparently simple system may be [50].

Coupling between different modes was revealed with acetylene adsorbed on Cu(1 0 0) [51]. Excitation of the C—H stretch mode at 358 meV by tunneling electrons caused a 10-fold increase of the rotation rate. Similar experiments were performed with other weakly held adsorbates at low temperatures [52,53].

In the latter case [53], ammonia molecules adsorbed on Cu(1 0 0) were activated by tunneling electrons either through the stretching vibration or the inversion vibration leading to translation or desorption, respectively.

Similar experiments with more tightly held adsorbates require higher bias voltages [54–59]. In this way, chlorobenzene adsorbed on Si(1 1 1) was subjected to the selective dissociation of C–Cl bonds, and it was concluded that a two-electron mechanism is operating that couples vibrational excitation and dissociative electron attachment processes. As can be seen from Fig. 4.9, the yield of desorption increases linearly with the electron current indicating a single-electron process, while for dissociation the yield increases with the second power.

With the system CO/Cu(1 1 1), tip-induced desorption was combined with femtosecond laser techniques to get closer insight into the dynamics of this process [60]. As depicted in Fig. 4.10,

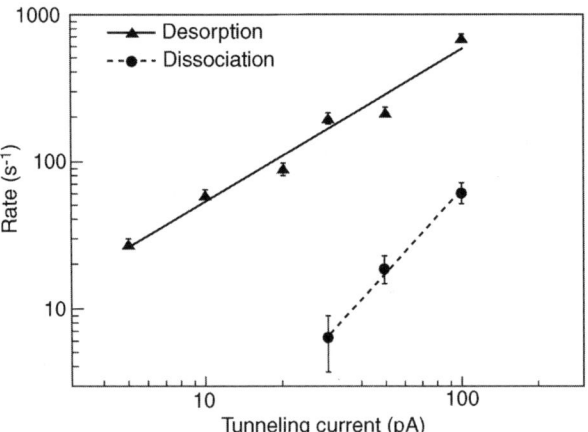

FIGURE 4.9. Rates for desorption and C–Cl bond dissociation for chlorobenzene adsorbed on Si(1 1 1) as a function of the current between the STM tip and the surface [59].

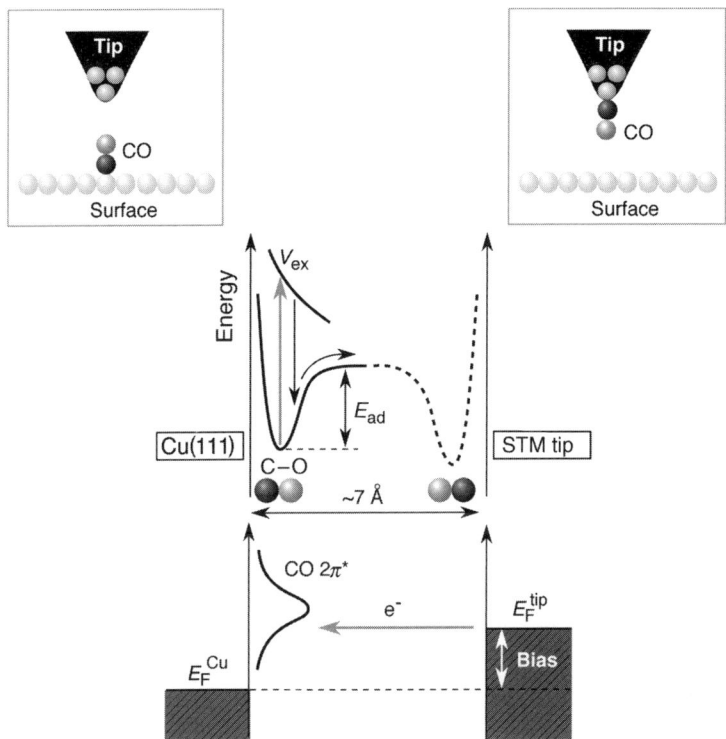

FIGURE 4.10. The mechanism of desorption of CO adsorbed on Cu(1 1 1) by electrons from an opposite STM tip by which the CO-$2\pi^*$-derived level is transiently populated [60].

electrons tunneling from the STM tip to an adsorbed CO molecule may cause their hopping from the surface to the tip if the bias voltage exceeds a threshold value of 2.4 eV. There is again a linear relation between desorption probability and tunneling current, indicating a single-electron mechanism. Probing the electronic density of states above the Fermi level by two-photon photoelectron spectroscopy revealed that this process involves the transient population of the CO-$2\pi^*$-derived level centered at 3.5 eV above E_F. This is associated with transition to a repulsive potential from which desorption through the DIET mechanism may take place. The desorption probability is then, however, only of the order 5×10^{-9} because of the very short lifetime of this excitation.

Time-resolved two-photon experiments revealed that the lifetime of an electron in the $2\pi^*$-derived level is indeed smaller than 5 fs, in agreement with the lifetimes of hot electrons in metals in this energy range [61]. It is believed that this marks the lower limit for the timescale of chemical reactions.

4.5. Surface Photochemistry

Associative desorption of hydrogen or the reaction between CO and O on a Ru(0 0 0 1) initiated by intense IR fs laser pulses, as outlined at the end of Chapter 3, are reactions caused by heating up the electron gas of the metal. The repetitive transition between ground-state and excited potentials in the DIMET process is responsible for the occurrence of the reaction that can still be considered as thermal since the electron gas equilibrates rapidly to an electron temperature and the process is still dominated by the ground-state potential. It is characteristic of this mechanism that the yield increases stronger than linear with the photon flux, reflecting its multiple excitation mechanism.

Irradiation of adsorbate-covered surfaces with higher energy photons (typically up to 6.4 eV) with lower intensities opens the possibility of direct valence excitation. Since the lifetimes of electronic excitations at metal surfaces are much shorter than those for nuclear motion, photochemical reactions appear rather improbable. Surprisingly, however, the cross sections determined for photodesorption were found to be comparable to those found for reactions with free molecules, mainly because the short lifetime of the excited state is compensated by a much larger cross section for absorption of the light [32,62–64]. This process takes place in the near-surface region of the metal (within about 10 nm), where relaxation of the photoexcited electrons leads to rapid establishment of a transient energy distribution. As depicted in Fig. 4.11, these hot electrons may scatter at the surface or are resonantly attached to an empty level of the adsorbate.

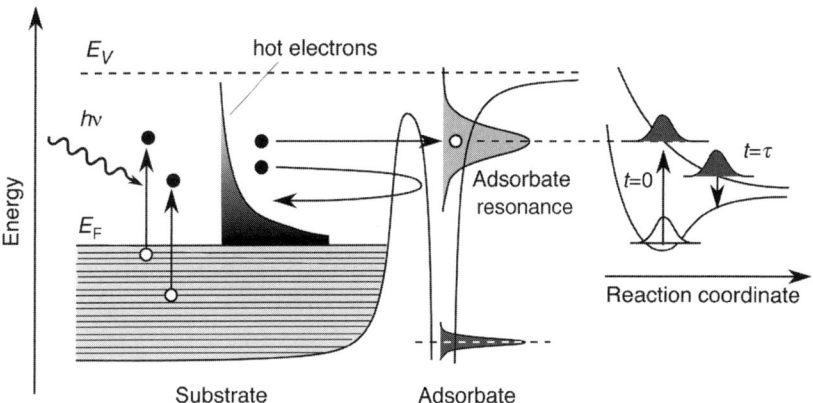

FIGURE 4.11. Energy diagram for an adsorbate covered metal surface under the influence of light absorption.

This implies a transition from the ground-state potential to that of an excited state. If the latter is repulsive, the nuclear motion becomes accelerated and may take up sufficient kinetic energy so that the particle can leave the surface after return to the ground state. This picture is equivalent to the MGR model of DIET, as discussed in Section 4.4.

Among the various systems studied in this way [36,62–64], in the following the photodesorption of ammonia from a Cu(1 1 1) surface will be discussed in more detail [65]. Up to a laser fluence of 8 mJ/cm^2, the desorption yield increases linearly with fluence, indicating that this process is initiated by single-photon absorption rather than by heating of the electron gas. Polarization measurements indicate an excitation process that is mediated by hot substrate electrons. Ammonia molecules come off the surface strongly peaked along the surface normal and with a mean translational energy of about 0.1 eV, which is much higher than the thermal energy corresponding to the surface temperature of 100K. Comparing NH_3 and ND_3 leads to a surprisingly high isotope effect of about 4. This suggests that the energy that is required to break the molecule–surface bond is acquired in an intramolecular (i.e., N−H) coordinate during the short-lived electronic excitation. For simulation of the desorption dynamics, model potential surfaces for the ground and excited states were proposed as a function of two coordinates, as reproduced in Fig. 4.12. The coordinate z refers to the ammonia center of mass to surface distance, and x denotes the distance of the plane of hydrogen atoms to the nitrogen atom. In the ground state, the molecule is bound to the surface through the N atom, and the H atoms are directed away from the surface. The excited state is

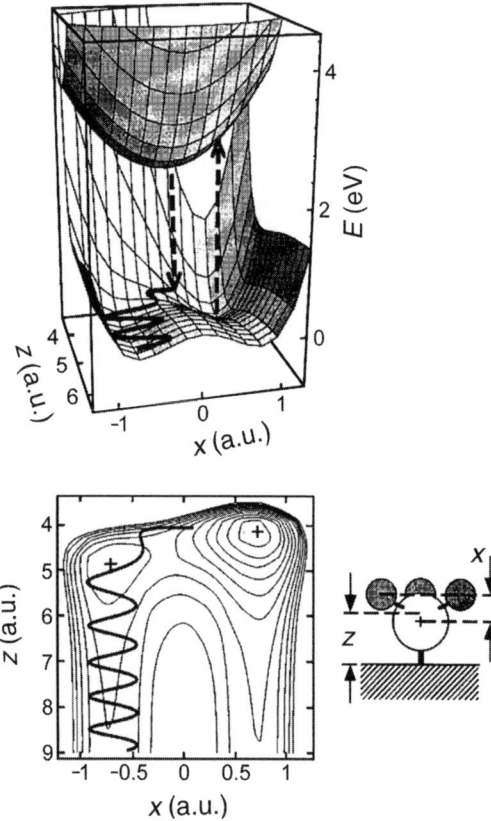

FIGURE 4.12. Photodesorption of ammonia from a Cu(1 1 1) surface. Model potential energy surfaces for the ground state and the excited state, respectively, as a function of the coordinates x and z. The lower part shows a contour plot of the ground-state PES and the solid line displays a typical desorption event on a calculated trajectory that had spent 9 fs on the excited-state PES [65].

characterized by a planar geometry, that is, a minimum of the PES at $x=0$. Upon excitation from the ground-state minimum (dashed line with up-arrow in Fig. 4.12), the system lands on a repulsive part of the excited-state potential, the gradient of which causes the plane of H atoms to bend down and the umbrella mode of the ammonia molecule becomes excited. After a few femtoseconds, the system will return to the ground state, but a small fraction will have picked up enough energy to desorb. One of the trajectories is plotted in the lower part of Fig. 4.12 and indicates that the molecule comes off the surface preferentially with the H atoms pointing down. Since the excitation of the N−H umbrella mode is decisive, the pronounced isotope effect becomes plausible. Further theoretical studies [66,67] essentially confirmed this concept. Determination of the vibrational-state distributions of the desorbing molecules provided information about the nonthermal population of states with different symmetry, which effects are sensitively affected by the potential for the inverted geometry [68].

Apart from desorption, photoexcitation may also cause dissociation of adsorbed molecules such as investigated with the system N_2O/Pt(1 1 1) [69]. Irradiation with 6.4 eV photons was found to cause desorption of N_2, N_2O, and O. The internal and translational energy distributions of the N_2 molecules suggest that they were completely thermalized before leaving the surface, while the desorbing N_2O molecules are detected with excess translational and vibrational energies.

Particles created by photodissociation may also move across the surface with enhanced reactivity. The formation of CO_2 upon irradiation of a $CO + O_2$ adlayer on Pt(1 1 1) was indeed the first hint for a "hot" adatom mechanism of this reaction, as discussed in Chapter 3 [70]. Subsequent detailed studies of this system [71] showed that the CO_2 molecules formed in this way at 25K substrate temperature are strongly peaked along the surface normal and exhibit translational energies up to 1.35 eV.

Photochemistry of adsorbed O_2 molecules results presumably from transient capture of an excited metal electron by the $3\sigma_u^*$ state of O_2. After decay to the ground state, the excited molecules may rapidly diffuse, desorb, or dissociate [72]. By monitoring the collision-induced desorption of coadsorbed noble gas atoms, a value of 0.7 eV for the kinetic energy of the fastest "hot" O adatoms was determined [73].

References

1. A. M. Wodtke, J. C. Tully, and D. J. Auerbach, *Int. Rev. Phys. Chem.* **23** (2004) 613.
2. C. Frischkorn and M. Wolf, *Chem. Rev.* **106** (2006) 4207.
3. L. König, I. Rabin, W. Schulze, and G. Ertl, *Science* **274** (1996) 1353.
4. I. Rabin, W. Schulze, and G. Ertl, *J. Chem. Phys.* **108** (1998) 5137.
5. I. Langmuir and K. H. Kingdon, *Phys. Rev.* **21** (1973) 381.
6. F. Haber and G. Just, *Ann. Phys.* **30** (1909) 411; **36** (1911) 308.
7. J. Kramer, *J. Phys.* **125** (1949) 739; **129** (1951) 34.
8. T. Greber, *Surf. Sci. Rep.* **28** (1997) 1.
9. T. Greber, K. Freihube, R. Grobecker, A. Böttcher, K. Hermann, G. Ertl, and D. Fick, *Phys. Rev. B* **50** (1994) 8755.
10. B. Kasemo, E. Törnqvist, J. K. Nørskov, and B. Lundqvist, *Surf. Sci.* **89** (1979) 554.
11. A. Böttcher, A. Morgante, T. Giessel, T. Greber, and G. Ertl, *Chem. Phys. Lett.* **231** (1994) 119.
12. K. Hermann, K. Freihube, T. Greber, A. Böttcher, R. Grobecker, D. Fick, and G. Ertl, *Surf. Sci.* **313** (1994) L806.
13. T. Greber, R. Grobecker, A. Morgante, A. Böttcher, and G. Ertl, *Phys. Rev. Lett.* **70** (1993) 1331.
14. L. D. Trowbridge and D. R. Herschbach, *J. Vac. Sci. Technol.* **18** (1989) 588.
15. E. B. de Blasi Bourdon and R. H. Prince, *Surf. Sci.* **144** (1984) 591.
16. L. Hellberg, J. Strömqvist, B. Kasemo, and B. Lundqvist, *Phys. Rev. Lett.* **74** (1995) 4772.
17. J. Bergström and C. Nordling, in: *Alpha-, Beta-, and Gamma-Ray Spectroscopy* (ed. K. Siegbahn), Vol. 2, North Holland, 1965.
18. E. Hasselbrink, *Curr. Opin. Solid State Mater. Sci.* **10** (2006) 192.

19. H. Nienhaus, H. S. Bergh, B. Gergen, A. Majumdar, W. H. Weinberg, and E. W. McFarland, *Phys. Rev. Lett.* **82** (1999) 446.
20. B. Gergen, H. Nienhaus, W. H. Weinberg, and E. W. McFarland, *Science* **294** (2001) 2521.
21. H. Nienhaus, *Surf. Sci. Rep.* **45** (2002) 1.
22. J. R. Trail, D. H. Bird, M. Persson, and S. Holloway, *J. Chem. Phys.* **119** (2003) 4539.
23. J. R. Trail, H. C. Graham, D. M. Bird, M. Persson, and S. Holloway, *Phys. Rev. Lett.* **88** (2002) 166802.
24. G. W. Gadzuk, *J. Phys. Chem. B* **106** (2002) 8265.
25. D. Krix, R. Nünthel, and H. Nienhaus, *Phys. Rev. B* **75** (2007) 073410.
26. S. Glass and H. Nienhaus, *Phys. Rev. Lett.* **93** (2004) 168302.
27. E. Mildner, E. Hasselbrink, and D. Diesing, *Chem. Phys. Lett.* **432** (2006) 137.
28. G. A. Somorjai, *Catal. Lett.* **101** (2005) 1.
29. X. Ji, A. Zuppero, J. M. Gidwami, and G. A. Somorjai, *Nano Lett.* **5** (2005) 753.
30. X. Ji, A. Zuppero, J. M. Gidwami, and G. A. Somorjai, *J. Am. Chem. Soc.* **127** (2005) 5792.
31. J. Y. Park and G. A. Somorjai, *Chemphyschem* **7** (2006) 1409.
32. J. Y. Park and G. A. Somorjai, *J. Vac. Sci. Technol. B* **24** (2006) 1967.
33. J. Y. Park, J. R. Renzas, A. M. Contreras, and G. A. Somorjai, *Top. Catal.* **46** (2007) 217.
34. J. Y. Park, J. R. Renzas, B. B. Hsu, and G. A. Somorjai, *J. Phys. Chem. C* **111** (2007) 15331.
35. R. D. Ramsier and J. T. Yates, *Surf. Sci. Rep.* **12** (1991) 243.
36. F. M. Zimmermann and W. Ho, *Surf. Sci. Rep.* **22** (1995) 127.
37. D. Menzel and R. Gomer, *J. Chem. Phys.* **41** (1964) 3311.
38. P. A. Redhead, *Can. J. Phys.* **42** (1964) 886.
39. P. R. Antoniewicz, *Phys. Rev. B* **21** (1980) 3811.
40. M. Akbulut, T. E. Madey, L. Parenteau, and L. Sanche, *J. Chem. Phys.* **105** (1996) 6032.
41. M. Akbulut, T. E. Madey, L. Parenteau, and L. Sanche, *J. Chem. Phys.* **105** (1996) 6043.
42. J. J. Czyzewski, T. E. Madey, and J. T. Yates, *Phys. Rev. Lett.* **32** (1974) 727.
43. M. J. Dresser, M. D. Alvey, and J. T. Yates, *Surf. Sci.* **169** (1986) 91.
44. M. D. Alvey and J. T. Yates, *J. Am. Chem. Soc.* **110** (1988) 1782.
45. E. Nitschke, G. Ertl, and J. Küppers, *J. Chem. Phys.* **74** (1981) 5911.

46. J. A. Stroscio and D. Eigler, *Science* **254** (1991) 1399.
47. P. H. Avouris, *Acc. Chem. Res.* **28** (1995) 95.
48. B. C. Stipe, M. A. Rezaei, W. Ho, S. Gao, M. Persson, and B. I. Lundqvist, *Phys. Rev. Lett.* **78** (1997) 4410.
49. B. C. Stipe, M. A. Rezaei, and W. Ho, *Science* **27** (1998) 1907.
50. W. Ho, *Acc. Chem. Res.* **31** (1998) 567.
51. B. C. Stipe, M. A. Rezaei, and W. Ho, *Phys. Rev. Lett.* **81** (1998) 1263.
52. T. Komeda, Y. Kim, M. Kawai, B. N. J. Persson, and H. Ueba, *Science* **195** (2002) 2055.
53. J. I. Pascual, N. Lorente, Z. Song, H. Conrad, and H. P. Rust, *Nature* **423** (2003) 525.
54. F. W. Fishlock, A. Oral, R. G. Egdelt, and J. B. Pethica, *Nature* **404** (2000) 743.
55. L. J. Lauhou and W. Ho, *Phys. Rev. Lett.* **84** (2000) 1527.
56. S. W. Hla, L. Bartels, G. Meyer, and K. H. Rieder, *Phys. Rev. Lett.* **85** (2000) 279.
57. P. H. Lu, J. C. Polanyi, and D. Rogers, *J. Chem. Phys.* **111** (1999) 9905.
58. L. Soukiassan, A. J. Mayne, M. Carbone, and G. Dujardin, *Phys. Rev. B* **68** (2003) 03503.
59. P. A. Sloan and R. E. Palmer, *Nature* **434** (2005) 367.
60. L. Bartels, G. Meyer, K. H. Rieder, D. Velic, E. Knoesel, A. Hotzel, M. Wolf, and G. Ertl, *Phys. Rev. Lett.* **80** (1998) 2004.
61. E. Knoesel, A. Hotzel, T. Hertel, M. Wolf, and G. Ertl, *Surf. Sci.* **368** (1996) 76.
62. X. L. Zhou, X. Y. Zhu, and J. M. White, *Surf. Sci. Rep.* **13** (1991) 73.
63. H. L. Dai and W. Ho (eds.), *Laser Spectroscopy and Photochemistry at Metal Surfaces*, Vols. I and II, World Scientific, Singapore, 1995.
64. H. Petek and S. Ogawa, *Prog. Surf. Sci.* **56** (1998) 239.
65. T. Hertel, M. Wolf, and G. Ertl, *J. Chem. Phys.* **102** (1995) 3414.
66. E. Hasselbrink, M. Wolf, S. Holloway, and P. Saalfrank, *Surf. Sci.* **263** (1996) 179.
67. H. Guo and T. Seideman, *J. Chem. Phys.* **103** (1995) 9062.
68. K. H. Bornscheuer, W. Nessler, M. Binetti, E. Hasselbrink, and P. Saalfrank, *Phys. Rev. Lett.* **78** (1997) 1174.
69. D. P. Masson, E. J. Lanzendorf, and A. C. Kummel, *J. Chem. Phys.* **102** (1995) 9096.
70. W. D. Mieher and W. Ho, *J. Chem. Phys.* **91** (1989) 4755.
71. V. A. Ukraintsev and I. Harrison, *J. Chem. Phys.* **96** (1992) 6307.

References

72. I. Harrison, *Acc. Chem. Res.* **31** (1998) 631.
73. A. N. Artsynkhovich and I. Harrison, *Surf. Sci.* **350** (1966) L199.
74. J. D. White, J. Chen, D. Matsiev, D. J. Auerbach, and A. M. Wodtke, *Nature* **433** (2005) 503.
75. N. H. Nahler, J. D. White, J. LaRue, D. J. Auerbach, and A. M. Wodtke, *Science* **321** (2008) 1191.

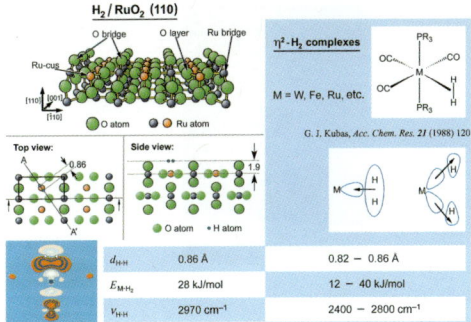

FIGURE 1.6. The bonding of H_2 on a single Ru atom or on a $RuO_2(1\,1\,0)$ surface [26].

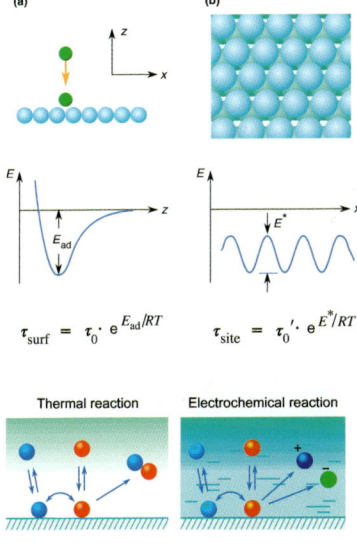

FIGURE 1.9. Potential of a chemisorbed particle along a certain direction on a single-crystal surface: (a) lifetime against desorption τ_{surf} determined by the adsorption energy E_{ad}; (b) surface residence time for motion τ_{site} determined by the activation energy for diffusion E^*.

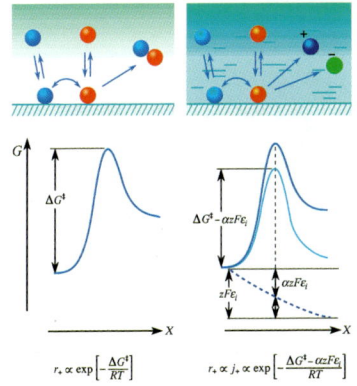

FIGURE 2.12. Energetics and kinetics of a thermal and an electrochemical reaction.

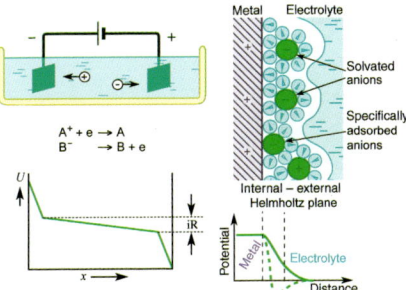

FIGURE 2.13. Potential distribution in an electrochemical cell (left) and structure and potential of the double layer (right).

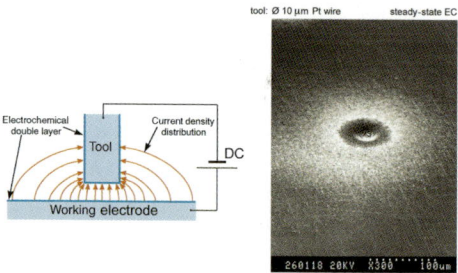

FIGURE 2.14. Electrochemical surface machining: current density distribution (left) and resulting hole (right).

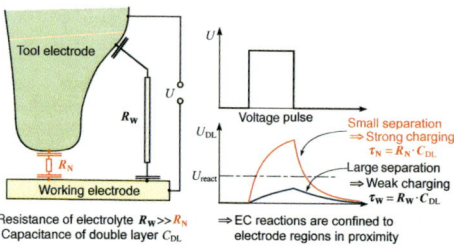

FIGURE 2.15. Principle of electrochemical micromachining by short voltage pulses [14].

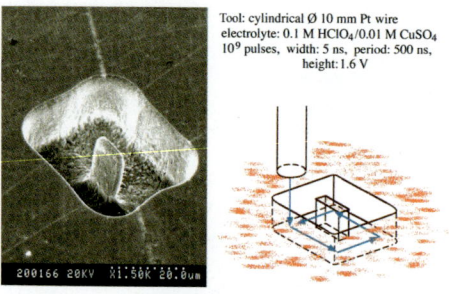

FIGURE 2.16. Electrochemical creation of a microstructure on a Cu surface [14].

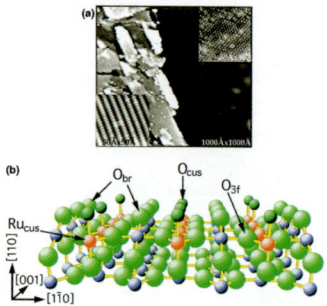

FIGURE 2.26. Structural transformation of the Ru(0 0 0 1) surface into a $RuO_2(1\,1\,0)$ overlayer under the influence of oxygen: (a) STM image exhibiting both phases; (b) ball model of the $RuO_2(1\,1\,0)$ surface.

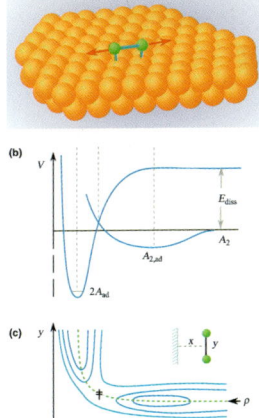

FIGURE 3.3. Energetics of dissociative adsorption of a diatomic molecule. (a) Schematic cartoon. (b) One-dimensional Lennard–Jones potential. (c) Two-dimensional representation with contour lines as a function of the distance x from the surface and the separation y between the two atoms.

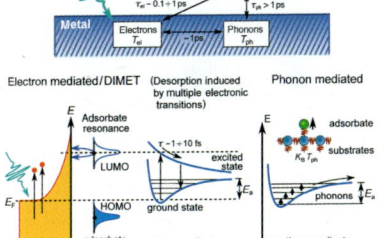

FIGURE 3.14. The mechanism of desorption caused by hot electrons (DIMET) created by absorption of a fs IR pulse versus phonon-mediated desorption from the ground state [51].

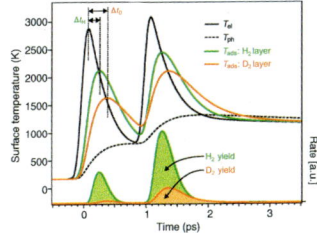

FIGURE 3.15. Variation of the electronic, phonon, and adsorbate temperatures (T_{el}, T_{ph}, and T_{ads}) with time following irradiation of H (D) covered Ru(0 0 0 1) surface by two subsequent fs IR pulses [51].

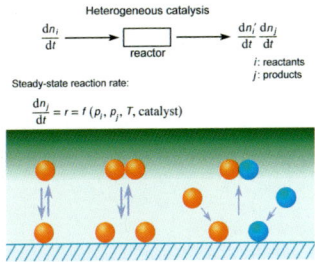

FIGURE 5.1. The principle of heterogeneous catalysis.

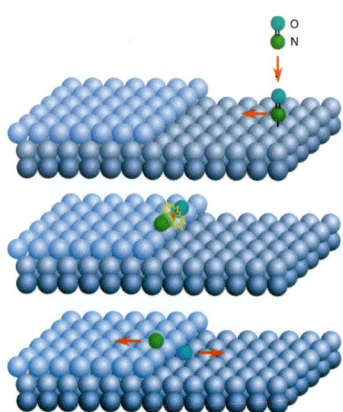

FIGURE 5.2. The role of terraces in the dissociation of NO at the steps of a Ru(0 0 0 1) surface [9].

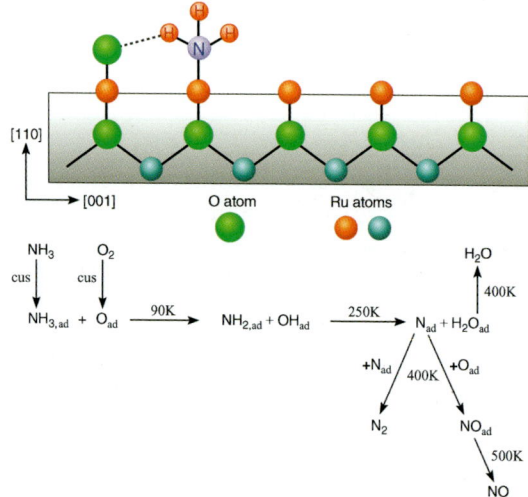

FIGURE 5.7. Mechanism of NH_3 oxidation on a RuO_2 (1 1 0) surface [34].

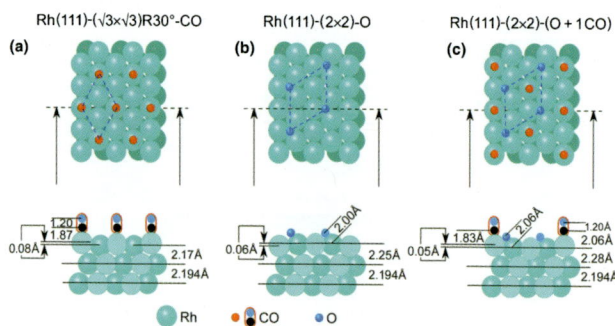

FIGURE 6.15. Ordered structures formed on a Rh(1 1 1) surface [56]. (a) $\sqrt{3}$ phase of CO. (b) 2×2 phase by O. (c) 2×2 phase by coadsorption of O + CO.

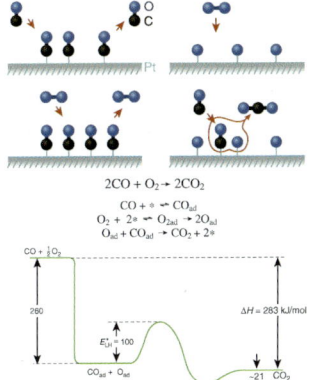

FIGURE 6.16. Mechanism and potential diagram for catalytic CO oxidation at a Pt(1 1 1) surface at low coverages (energies in kJ/mol).

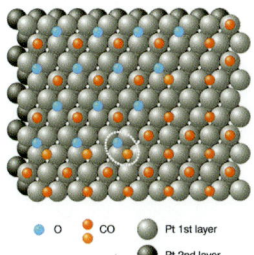

FIGURE 6.18. Cartoon illustrating the boundary between the mixed 2×2 (O + CO) phase and the $c2 \times 4$ phase of CO on Pt(1 1 1).

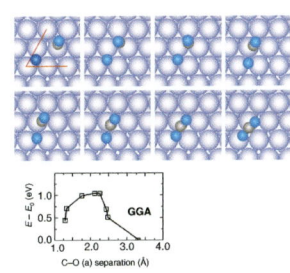

FIGURE 6.19. DFT calculations for the energy on a Pt(1 1 1) surface if O_{ad} and CO_{ad} approach each other to form CO_2 [60].

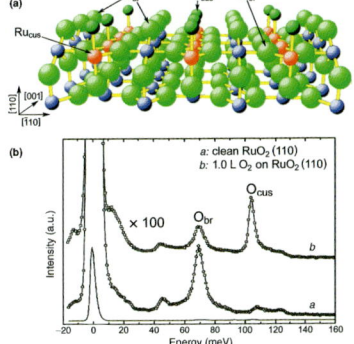

FIGURE 6.21. The RuO_2(1 1 0) surface [69]. (a) Ball model with additional O atoms adsorbed on cus sites. (b) Corresponding vibrational spectrum.

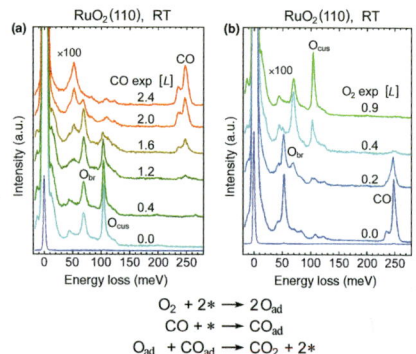

$$O_2 + 2* \rightarrow 2O_{ad}$$
$$CO + * \rightarrow CO_{ad}$$
$$O_{ad} + CO_{ad} \rightarrow CO_2 + 2*$$

FIGURE 6.22. Vibrational spectra from a $RuO_2(1\,1\,0)$ surface showing the progress of the reaction between adsorbed O and CO.

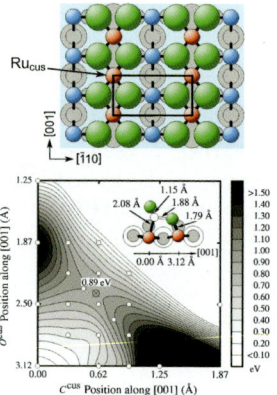

FIGURE 6.23. Theoretical energy diagram for the interaction between O and CO on a $RuO_2(1\,1\,0)$ surface [70].

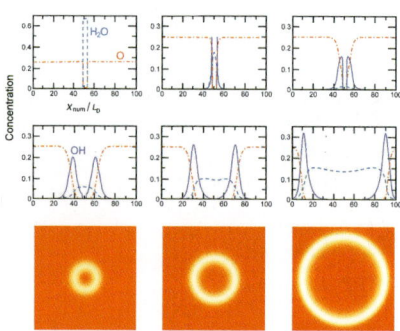

FIGURE 6.29. Numerical simulation of the formation and propagation of an OH_{ad} wave in the $H_2 + O_2$ reaction on $Pt(1\,1\,1)$ [85,86].

FIGURE 7.2. Variation with time of the number of furs (in thousands) from hares and lynxes delivered to the Hudson's Bay Company [18].

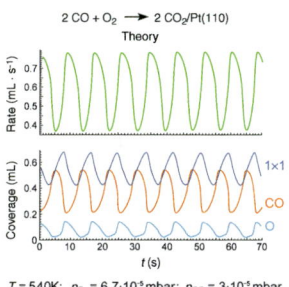

FIGURE 7.7. Solution of the set of ordinary differential equations presented in Fig. 7.6 and describing the kinetic oscillations in CO oxidation on Pt(1 1 0) [25].

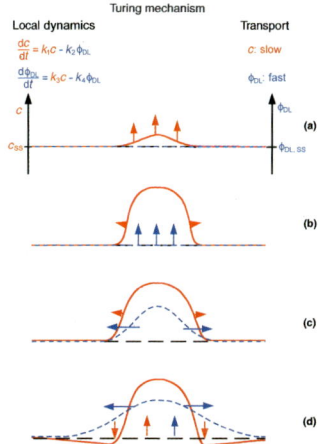

FIGURE 8.1. Principle of the mechanism underlying the formation of electrochemical Turing patterns [5].

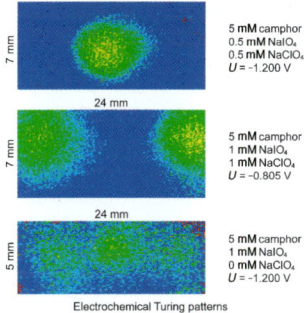

FIGURE 8.2. Some examples for the formation of electrochemical Turing patterns on a thin Au(1 1 1) film [12].

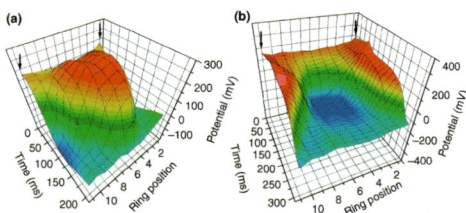

FIGURE 8.4. Space–time plots of the potential distributions on a Pt ring electrode during passive–active transitions. The ring position refers to Fig. 8.3 [44] (a) Local triggering. (b) Remote triggering.

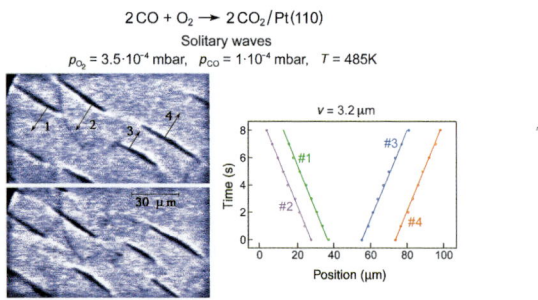

FIGURE 8.7. Solitary O waves propagating along the [0 0 1] direction on a Pt (1 1 0) surface during CO oxidation [21].

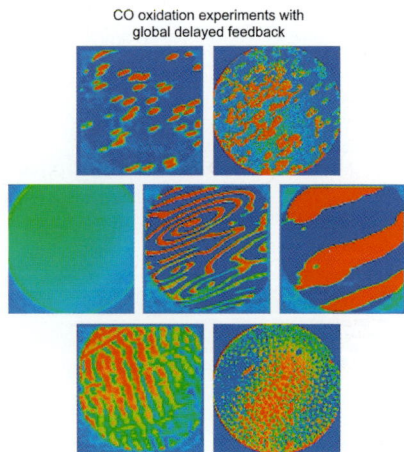

FIGURE 8.17. Summary of controlled formation of patterns in CO oxidation on Pt(1 1 0) by global delayed feedback with varying delay times and strengths of feedback [42].

CHAPTER 5

Principles of Heterogeneous Catalysis

5.1. Introduction

Heterogeneous catalysis is the most important manifestation of reactions at solid surfaces [1–6]. About 90% of all industrial chemical processes involve heterogeneous catalysis, which is not only the basis of chemical and petroleum industries but is also of crucial significance for protecting the environment and for solution of the energy problem and so on. The principle of this phenomenon is depicted in Fig. 5.1: The catalyst is inside a reactor that is typically operated in continuous flow, so the steady-state reaction rate is determined by the external parameters, partial pressures, temperature, and flow rate, and by the nature of the catalyst that exposes its surface to the molecules taking part in the reaction. The elementary processes involved are depicted in the lower part of Fig. 5.1 and comprise both dissociative and nondissociative adsorption and desorption steps, as well as surface diffusion of the adsorbates and their surface reaction to product molecules that then desorb. As already outlined in Section 1.1, a typical catalyst consists of nanometer-sized particles with a specific surface of up to $1000 \, m^2/g$ and exposing different crystal faces

---

Reactions at Solid Surfaces. By Gerhard Ertl
Copyright © 2009 John Wiley & Sons, Inc.

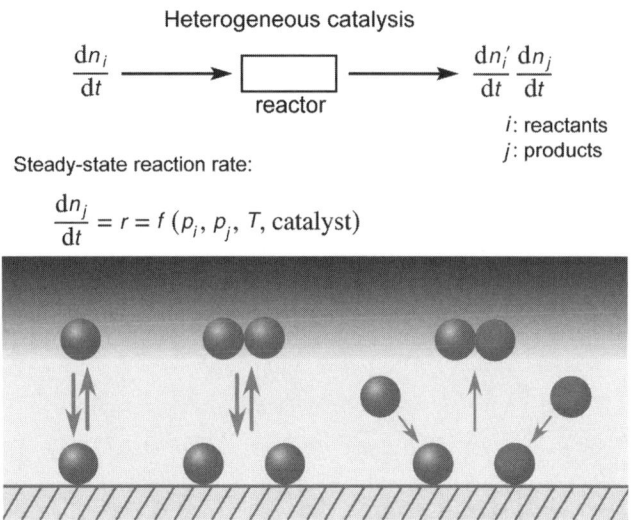

FIGURE 5.1. The principle of heterogeneous catalysis. (See color insert.)

with various structural defects. If the catalyst consists of more than one component (which is often the case), the surface composition will usually be different from that of the bulk and may even differ between the different crystal planes. All these complications together with the fact that under the conditions of "real" catalysts the actual nature of the surface under working conditions is unknown rendered this field for long times the character of a "black art." As a consequence industrial catalysis is even at present largely governed by empirical experience. However, the development of the "surface science" approach, as sketched in Section 1.1, enabled in the meantime closer insights into the elementary steps involved and direct catalysis research toward real science.

The use of single-crystal surfaces under ultrahigh vacuum conditions as model systems will necessarily differ substantially from the conditions of real catalysis. These materials and pressure gaps can be bridged along various ways:

(i) Systematic studies with different single-crystal surfaces, including well-defined defects such as monoatomic steps, may be compared with experiments using varying particle size, since with the latter the relative portions of exposed crystal planes will systematically vary [7].

(ii) Analysis of the actual composition of a catalyst (as close as possible to its working conditions) enables deliberate modification of the chemical composition of the model surface. Very often the surface of a catalyst will change its structure and composition under working conditions, that is, the reaction "digs its own bed."

(iii) The pressure gap is often considered to present the most serious problem. If it is found, the problem will often be a consequence of the materials gap, as will be shown later for the case of CO oxidation on Ru catalysts. In fact, the kinetics of a catalytic reaction will be governed by the surface concentrations of the species involved rather than by the partial pressures in the gas phase. Higher coverages, however, can also be achieved by lowering the temperature instead of rising the pressure. The steady-state surface concentration of a certain species will generally be determined by the kinetics of parallel, consecutive, or competitive reaction steps and may thus be rather low even if the overall reaction is operated at high pressures. Information about the reaction mechanism may further be obtained by studying the backward reaction (at low pressure) since this proceeds through the same intermediate microscopic steps.

5.2. Active Sites

The term "active sites" was introduced by Taylor [8] in 1925 in the following way: "A surface may be regarded as composed of atoms in varied degrees of saturation by neighbouring metal atoms. The varying degree of saturation of the catalyst atoms also involves a varying catalytic capacity of the surface atoms. There will be all extremes between the case in which all the atoms in the surface are active and that in which relatively few are so active."

This statement just expresses the influence of the surface structure on its reactivity, as was discussed in Section 2.1. The STM images of Fig. 2.4 on the dissociative adsorption of NO on a Ru(0 0 0 1) surface with monoatomic steps are a direct experimental demonstration of the enhanced reactivity of step atoms [9]. These findings become plausible in view of the activation energies for dissociation at various surface sites as derived from DFT calculations [10]: While this amounts to 1.28 eV at a flat terrace site, it is only 0.15 eV at a step site. Since reaction rates (i.e., sticking coefficients) exponentially depend on activation energies, it becomes obvious that in this case dissociation occurs exclusively at the step sites. Surprisingly, the macroscopic sticking coefficient revealed to be essentially independent of the step density. As illustrated in Fig. 5.2, the NO molecules impinging on the terrace sites are (weakly) adsorbed and diffuse rapidly across the surface until they hit a step where they dissociate. The wider the terrace, the more molecules will reach this state as long as their mean diffusion length prior to desorption exceeds the distance between the neighboring steps. In this way, a higher step concentration is compensated and, in fact, the whole surface becomes involved in the reaction.

Figure 5.3 depicts an Arrhenius diagram for the sticking coefficient of dissociative adsorption of N_2 on Ru(0 0 0 1) surfaces. The data for clean Ru(0 0 0 1) surfaces from two different laboratories [11,12] are in agreement with each other, although the samples were unlikely to exhibit the same defect densities. The second set of data was obtained with a surface onto which a small amount of gold atoms had been deposited [12]. Now obviously all step sites are blocked, the sticking coefficient drops

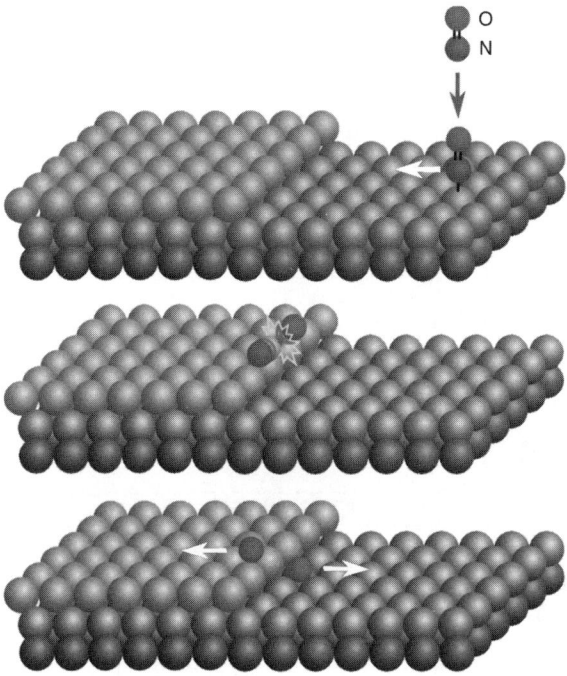

FIGURE 5.2. The role of terraces in the dissociation of NO at the steps of a Ru(0 0 0 1) surface [9]. (See color insert.)

by several orders of magnitude, and the activation energy rises from 0.4 to 1.3 eV, the latter value being characteristic of the terrace sites. Step sites exhibiting a higher degree of unsaturated valencies are the preferred location for the adsorption of foreign atoms that block the catalytically active centers. This becomes, for example, evident from Fig. 5.4 that shows the STM image from a Al(1 1 1) surface onto which a small concentration of S atoms had been deposited that preferentially adsorb at the steps. This also explains why often the small amounts of foreign atoms kill the catalytic activity and act as poisons.

These examples illustrate not only the role of active sites but also how complex their role is for macroscopic kinetics. This leads us to the general problem of how to define catalytic activity.

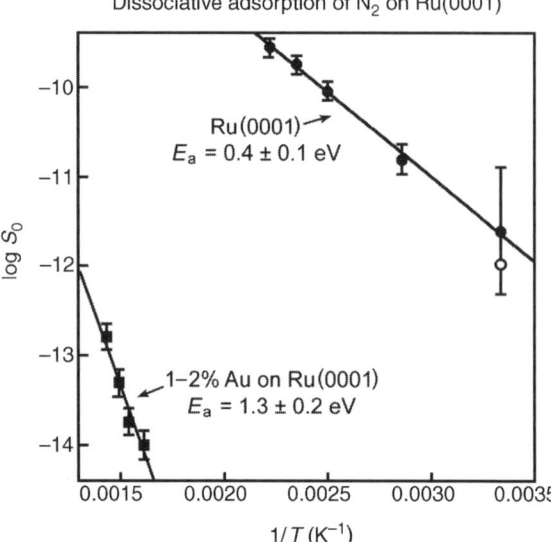

FIGURE 5.3. Arrhenius diagram (log s_0 versus $1/T$) for the sticking coefficient for dissociative nitrogen surface at a Ru(0 0 0 1) surface, and the influence of a small concentration of Au atoms blocking the "active sites" at the steps [12].

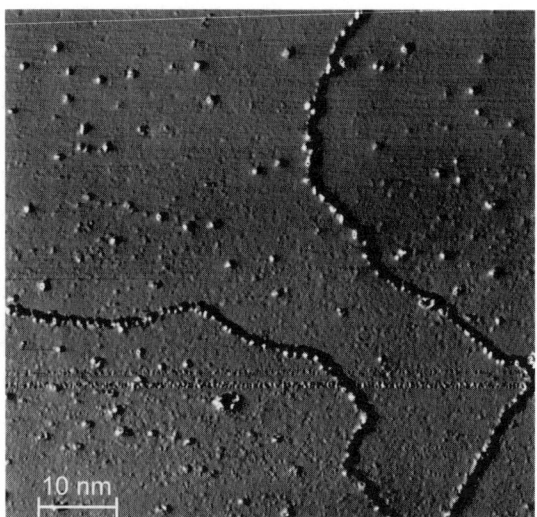

FIGURE 5.4. STM image from a Al(1 1 1) surface with a small concentration of S atoms that preferentially segregate to the steps.

Sometimes it is given as reaction rate per unit weight or specific surface area of the catalyst, but more instructive is its definition in terms of the turnover frequency (TOF). This is defined as the number of reaction events per active site and unit time and relies on the possibility to evaluate the correct density of active sites.

5.3. LANGMUIR–HINSHELWOOD VERSUS ELEY–RIDEAL MECHANISM

Among the elementary processes governing a heterogeneously catalyzed reaction as depicted in Fig. 5.1, the steps involving dissociative or nondissociative adsorption/desorption have already been discussed in detail in the preceding chapters. An additional aspect comes into play with regard to the reaction between two different species A + B to a product molecule. Two limiting cases for this process are discussed: In the Langmuir–Hinshelwood (LH) mechanism [13], both species are in the adsorbed state and fully thermally equilibrated with the surface, while with the Eley–Rideal (ER) mechanism [14], an adsorbed particle A reacts directly by collision of an impinging species B.

Generally, almost all catalytic surface reactions proceed via the LH mechanism. Experimental verification was, for example, obtained in studies in which after adsorption of the reactants A + B, the gas phase is pumped off and the sample temperature continuously increased until the product molecules come off. This procedure is denoted as temperature-programmed reaction spectroscopy (TPRS). In particular, operation of the LH mechanism may be concluded if the relaxation time for product formation is long enough (>10 ps) to reach thermal equilibrium of the adsorbates with the surface. For the CO + O reaction on a Pd (1 1 1) surface, this delay time between the adsorption and product formation was determined in molecular beam experiments down to about 10^{-4} s, leading to derivation of the action energy for the LH step [15].

Reactions involving hot adatoms, such as discussed earlier, clearly represent an intermediate situation between the two limiting cases LH and ER, since the reactants are not in complete thermal equilibrium with the surface (as for LH). Neither does the reaction take place by direct collision from the gas phase (as for ER). Such a kind of intermediate mechanism ("precursor" mechanism) has been proposed [16], whereafter one of the reactants is not fully accommodated with the surface but reacts from a state that is vibrationally excited with respect to motion normal to the surface.

Clear cases for the operation of the genuine ER mechanism are still rather scarce, and experimental evidence is so far restricted to reactions involving the impact of *atoms* (i.e., energetic species) on the surface.

First experimental evidence for the operation of non-LH behavior was found with CO oxidation on Pt(1 1 1), when O *atoms* together with CO molecules were striking the surface [17]. If the energy content of the product molecules was probed by IR chemiluminescence, CO_2 molecules formed by impact of O atoms exhibited a higher degree of internal excitation than when O_2 molecules were employed, suggesting some kind of "memory" of the initial state of the reactants [18]. The suggested operation of the ER mechanism was supported by classical trajectory calculations [19], whereafter the most reactive events occur within the first few picoseconds after the impact. However, a substantial fraction of product molecules is only formed after somewhat longer times, which are therefore more appropriately classified as hot adatom events.

The most detailed experiments on the operation of the ER mechanism were performed with reactions involving the impact of hydrogen atoms, such as H + D_{ad} → HD. The energy balance in this case reads $E_{HD} = E_{diss}$ (HD) + E_{kin} (H) − E_{ad} (D), which amounts to about 2.3 eV for Cu(1 1 1). Such an amount of energy carried away by the product molecules in their translational, vibrational, and rotational degrees of freedom was indeed found in the state-resolved measurements [20,21], and the conclusion of the ER mechanism was supported by theory [22,23].

However, experiments with the $H + D_{ad}$ reaction on Ni(1 1 0) revealed, apart from HD, also D_2 as product [24]. In addition, the rate of HD formation was not found to be simply proportional to the D_{ad} coverage as expected for a pure ER mechanism [25,26]. As a consequence, a more complex sequence of elementary steps was proposed [25]: Apart from the pure ER channel, a fraction of the incident H atoms may become adsorbed or trapped (without full accommodation) as hot adatoms. The latter may then react with or transfer their energy to adsorbed D atoms that thus become excited and may even react further to D_2. Computer simulations were able to qualitatively rationalize the observed experimental observations [27,28].

A general conclusion is that the pure ER mechanism barely operates, not even in reactions with atomic species, and that the actual situation is more complex and will presumably involve partly accommodated "hot" adparticles.

5.4. COADSORPTION

Since we have to conclude that catalytic reactions proceed essentially along the Langmuir–Hinshelwood mechanism, the situation that two different kinds of adsorbed species are present on the surface has to be taken into account. Although pronounced interactions between different species may exist, these will retain their molecular identity as can be probed by surface spectroscopic techniques.

If two kinds of particles are present on the surface and if we assume that their mutual interactions act pairwise and independently (which is a crude approximation), then the interaction energies ε_{AA}, ε_{AB}, and ε_{BB} lead to the following limiting situations that can be experimentally distinguished, for example, by STM or LEED [29].

(i) If $\varepsilon_{AA} + \varepsilon_{BB} - 2\varepsilon_{AB} < 0$, the two species will effectively repel each other, so the surface consists of two phases, namely, domains consisting either of pure A or pure B. This case of *competitive adsorption* is schematically sketched in Fig. 5.5a. Only that fraction of the surface that is not already covered by A may be occupied by B particles. The reaction between A and B will then be restricted to the boundaries between the two phases.

(ii) If $\varepsilon_{AA} + \varepsilon_{BB} - 2\varepsilon_{AB} > 0$, there will be an effective attraction between A and B (or less repulsion between equal particles) and a mixed phase, as sketched in Fig. 5.5b, will be formed. This situation is denoted as *cooperative adsorption*. A and B are then in intimate contact, but not randomly distributed.

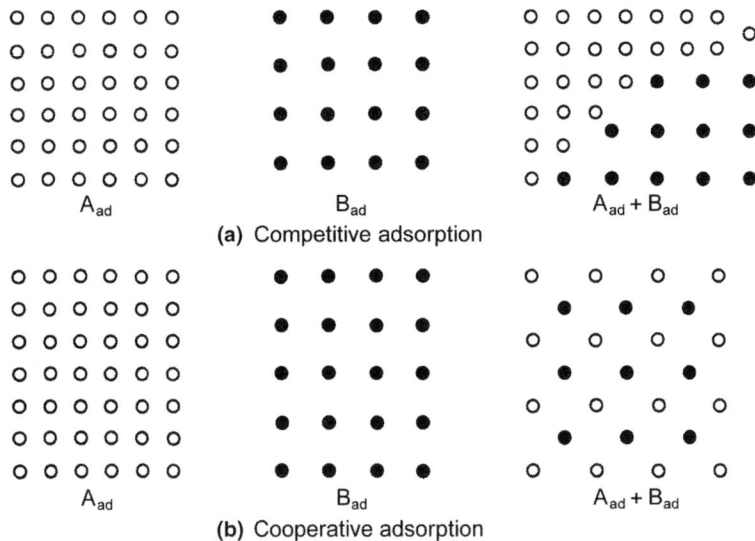

FIGURE 5.5. Schematic surface configuration of two species A and B [29]. (a) Competitive adsorption. (b) Cooperative adsorption.

These are in fact only limiting cases, and real systems will be even more complex. With increasing temperature, the degree of disorder will also increase, eventually leading to random distribution. In addition, new phases may be formed with variation of the composition of the overlayer.

Apart from these effects on the surface order, other consequences such as changes of the effective adsorption strength or displacement onto different adsorption sites as well as modification of the electronic properties may occur, so the actual situation may become rather complex.

The consequences for the kinetics of adsorption and desorption are quite obvious. The occupation of adsorption sites by a second species will usually reduce the sticking coefficient. This may lead to the effect that under steady-state conditions, the more strongly held adsorbate is not necessarily most abundant on the surface, as will become obvious in Section 6.3, with the oxidation of CO on Pt. A second species may, however, also increase the sticking coefficient and therefore acts as *promoter*. This will be outlined in Section 6.1 with the effect of K on the sticking coefficient for nitrogen in catalytic ammonia synthesis.

5.5. KINETICS OF CATALYTIC REACTIONS

As illustrated in Fig. 5.1, the steady state of a catalytic reaction depends (apart from transport processes) on the external parameters, temperature T and partial pressures of reactants and products, p_i, p_j, respectively. These parameters determine the surface concentrations of the reaction intermediates, which in turn are governed by the overall reaction mechanism. Description of the reaction rate r depending on the external parameters is achieved on various levels:

(a) Purely empirical *power laws* of the form $r = k \cdot p_A^a \cdot p_B^b$ may be used to adjust the unknown parameters k, a, and b to some experimental data, from where extrapolation to (not too different) other conditions may be achieved. This approach is of some use for practical applications, but contains no information about the progress of the reaction.

(b) Most common is what we shall call the *Langmuir approach* that is the most frequently applied attempt to describe the kinetics of a catalytic reaction. It starts with the assumption of a possible reaction mechanism, possibly supported by various experimental data on reaction intermediates. The various steps are then formulated in terms of rate equations for adsorption, desorption, and surface reaction steps. If adsorption/desorption steps are fast compared to the rate of surface reaction, these can be considered to be in equilibrium and the coverages for the different surface species are related with the respective partial pressures through adsorption isotherms. The latter should in principle be carefully determined experimentally, since even for a single component the adsorption isotherm is usually complex due to effects of the coverage on the kinetics and energetics of adsorption even with a uniform single-crystal surface, apart from the additional complications due to coadsorption as discussed above. If, finally, the "real" surfaces are considered with the various structural elements, the situation looks rather hopeless. Nevertheless, the Langmuir adsorption isotherm with its crude underlying assumptions may provide rather satisfactory results, probably because several effects are able to compensate each other. Dumesic et al. [30] argue that the actual presence of different surface sites and the effects of surface coverage may well be responsible for the robustness of a catalyst for operation over a wide range of reaction conditions. At a certain set of reaction conditions, an optimal set of surface sites will dominate the reactivity that may change into another subset if the temperature and pressure conditions are varied.

If we return to the simple Langmuir picture, the rate of a surface reaction of the type

$$A_{ad} + B_{ad} \xrightarrow{k} C \tag{5.1}$$

is then formulated as $r = k\Theta_A \Theta_B$, where the fractional coverages Θ_A and Θ_B are related with the respective partial pressures p_A and p_B through

$$\Theta_A = \frac{b_A p_A}{1 + b_A p_A + b_B p_B} \tag{5.2}$$

$$\Theta_B = \frac{b_B p_B}{1 + b_A p_A + b_B p_B} \tag{5.3}$$

This modification of the original Langmuir isotherm is based on the assumption that an adsorption site may either be occupied by A or B and that the product molecule is so weakly adsorbed that its surface concentration is negligible. Equation 5.1 then becomes

$$r = k\Theta_A \Theta_B = \frac{b_A b_B p_A p_B}{(1 + b_A p_A + b_B p_B)^2} \tag{5.4}$$

For constant T and p_B, the rate will pass through a maximum upon variation of p_A that is reached when $\Theta_A = \Theta_B$. If $b_A p_A \ll b_B p_B + 1$, the rate will increase linearly with p_B, while in the opposite case $r \approx (b_B p_B / b_A p_A)$, that is, further increase of p_A lowers the rate since the adsorption of A blocks sites for adsorption of B. In Section 6.3, the kinetics of CO oxidation on a $RuO_2(0\,0\,0\,1)$ surface will be described, which approximately shows such a behavior.

(c) The next step is denoted as *microkinetics*. The kinetics is now based on detailed information about the reaction mechanism and about the rates of the elementary steps involved [31]. These input parameters are obtained from experiments with real catalysts or with single-crystal surfaces, as well as from DFT calculations. Modeling is then again performed on the basis of the Langmuir model in which the surface concentrations are treated as continuum variables in a mean field approximation. A compilation of systems that were treated in this way can be found in Ref. [31], and in Chapter 6 the application of this approach to a few reactions for which the mechanism is known will be discussed in more detail.

(d) The *kinetic Monte Carlo* approach takes into account that the actual neighborhood of an adsorbed particle is affecting its energetics and kinetics, which effect is not taken into consideration with the mean field approximation. There the adsorbed particles are placed on a lattice and randomly selected for determination of the energetics by summing up the interactions with its neighbors from where the probability for a reactive event is calculated. This approach provides detailed insights but is associated with substantial computational efforts. If the temperature is high enough so that diffusion is fast and the adsorbed particles more or less randomly distributed, a mean field approximation will become sufficiently appropriate [32,33].

At the end of this section we conclude that modeling of the kinetics of a catalytic reaction essentially relies on the Langmuir concept, assuming uniform surfaces with periodic arrangements of equal adsorption sites without interactions between the adsorbed particles and a "hit and stick" model for the adsorption process. The microscopic observations as discussed in the preceding chapters, however, reveal that none of these assumptions is fulfilled. Instead there exists (even with a single-crystal surface) an intrinsic heterogeneity as manifested by the existence of active sites, there exist different adsorption sites and pronounced interactions between adsorbed particles, and the non-instantaneous energy release may cause the transient existence of "hot" adparticles. Nevertheless, the Langmuir approach leads to satisfactory results if properly modified and applied only over a limited range of parameters, as will become evident with the examples discussed in Chapter 6.

5.6. SELECTIVITY

A catalyst often creates more than one product, and the preferred formation of the desired molecule becomes more important than the overall reactivity. In principle, two different types of reaction sequences have to be considered.

(i) Consecutive reactions of the type A → B → C. If B is the desired product, its concentration in a closed reactor passes through a maximum at which point the reaction has to be interrupted.
(ii) More relevant are parallel reactions of the type $A+B \begin{smallmatrix} r_1 \nearrow C \\ r_2 \searrow D \end{smallmatrix}$, where the selectivities for the formation of C and D may be formulated as $s_1 = r_1/(r_1 + r_2)$ and $s_2 = r_2/(r_1 + r_2)$, where r_1 and r_2 are the rates for the formation of the products C and D, respectively.

In a continuous flow reactor, these quantities will, of course, again depend on the external parameters p_i and T, as will be illustrated in the following with the example of ammonia oxidation on a $RuO_2(1\,1\,0)$ surface [34]. This Ostwald process is the basis for large-scale production of nitric acid that operates with platinum-based catalysts at temperatures >1000K where the rate is transport limited [35]. With the model system discussed here, the temperature is much lower (~500K) and the rate is determined by the surface chemistry.

As sketched in Fig. 5.6, reaction between NH_3 and O_2 leads either to NO or N_2, where NO is the desired product. Figure 5.6a shows the variation of the steady-state rates of NO and N_2 formation, respectively, as a function of the O_2 partial pressure at fixed temperature $T = 500K$ and NH_3 partial pressure of 1×10^{-7} mbar. r_{N_2} passes through a maximum, while r_{NO} continuously increases with increasing p_{O_2}. The respective selectivities are plotted in Fig. 5.6b: While s_{N_2} continuously decreases, s_{NO} continuously increases up to about 0.8. (At 550K, it comes even close to 1.)

The elementary steps involved, as determined by surface spectroscopic techniques, are depicted in Fig. 5.7. After adsorption

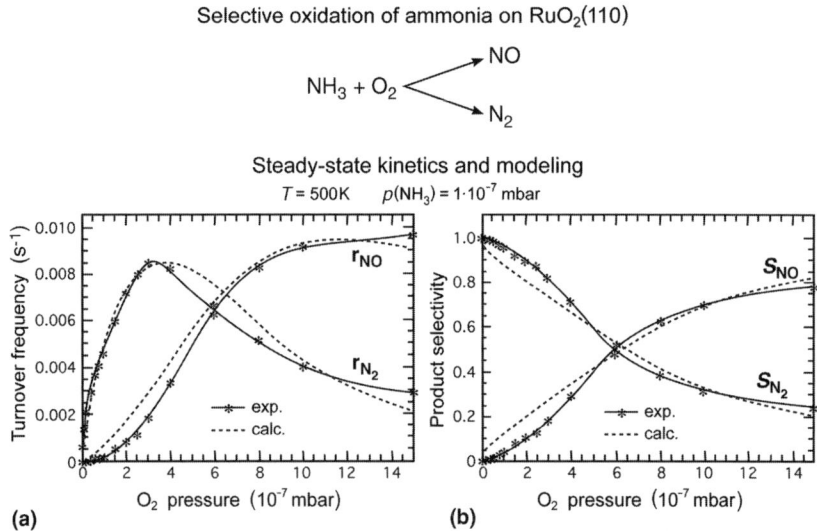

FIGURE 5.6. Catalytic oxidation of NH_3 on a $RuO_2(1\ 1\ 0)$ surface [34]. (a) Steady-state rates of N_2 and NO formation as a function of O_2 partial pressure with fixed T and p_{NH_3}. (b) Respective selectivities for the formation of N_2 and NO.

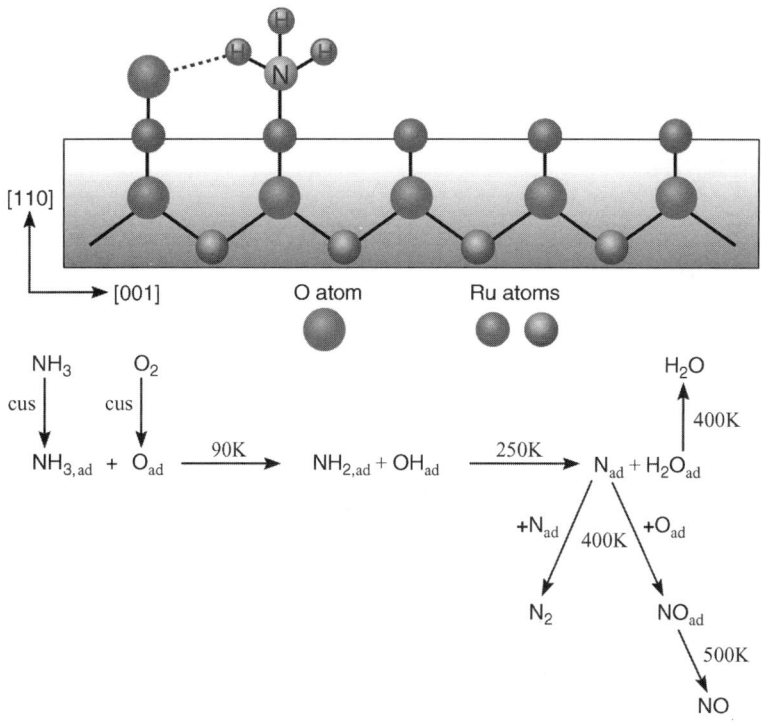

FIGURE 5.7. Mechanism of NH_3 oxidation on a RuO_2 (110) surface [34]. (See color insert.)

of the reactants, the first rapid step consists in the abstraction of one H atom from NH_3 by a neighboring O_{ad} atom through hydrogen bond interaction. From there onward, further dissociation of $NH_{2,ad}$ leads to N_{ad} that may either recombine with a neighboring O_{ad} to NO or with another N_{ad} to N_2. In this way it becomes plausible why even for N_2 formation O_2 is necessary and why the selectivity for NO formation increases with increasing O_2 partial pressure, where the probability that N_{ad} has O_{ad} as a neighbor to react with is enhanced.

Modeling of the kinetics along the strategy of Langmuir-type microkinetics yields the dotted lines in Fig. 5.6, where some of the rate parameters of the intermediate steps were fitted to provide good agreement with the experimental data.

References

1. I. Chorkendorff and J. W. Niemantsverdriet, *Concepts of Modern Catalysis and Kinetics*, Wiley-VCH, 2003.
2. J. M. Thomas and W. J. Thomas, *Principles and Practice of Heterogeneous Catalysis*, Wiley-VCH, Weinheim, 1997.
3. R. A. van Santen, P. W. N. H. v. Leewen, J. A. Moulijn, and B. A. Averill, *Catalysis: An Integrated Approach*, Elsevier, 1999.
4. G. Rothenberg, *Catalysis: Concepts and Green Applications*, Wiley-VCH, 2008.
5. G. Ertl, H. Knözinger, F. Schüth, and J. Weitkamp (eds.), *Handbook of Heterogeneous Catalysis*, Vol. **8**, Wiley-VCH, 2008.
6. R. A. van Santen and M. Neurock, *Molecular Heterogeneous Catalysis*, Wiley, 2006.
7. M. Boudart, *Adv. Catal.* **20** (1969) 153.
8. H. S. Taylor, *Proc. R. Soc. Lond. A* **108** (1925) 105.
9. T. Zambelli, J. Wintterlin, J. Trost, and G. Ertl, *Science* **273** (1996) 1688.
10. B. Hammer, *Phys. Rev. Lett.* **83** (1999) 3631.
11. H. Dietrich, P. Geng, K. Jacobi, and G. Ertl, *J. Chem. Phys.* **104** (1996) 1688.
12. S. Dahl, A. Logadottir, R. C. Egeberg, J. H. Larsen, I. Chorkendorff, E. Törnqvist, and J. K. Nørskov, *Phys. Rev. Lett.* **83** (1999) 1814.
13. (a) I. Langmuir, *Trans. Faraday Soc.* **17** (1922) 621; (b) C. N. Hinshelwood, *The Kinetics of Chemical Change*, Claredon Press, Oxford, 1940.
14. (a) E. K. Rideal, *Proc. Cambridge Phil. Soc.* **35** (1938) 130; (b) D. D. Eley, *Adv. Catal.* **1** (1948) 157; (c) D. D. Eley and E. K. Rideal, *Nature* **146** (1940) 401.
15. T. Engel and G. Ertl, *J. Chem. Phys.* **69** (1978) 1267.
16. J. Harris and B. Kasemo, *Surf. Sci.* **105** (1981) L281.
17. C. B. Mullins, C. T. Rettner, and D. J. Auerbach, *J. Chem. Phys.* **95** (1991) 8649.
18. C. Wei and G. L. Haller, *J. Chem. Phys.* **105** (1996) 810.
19. J. Ree, Y. H. Kim, and H. K. Shin, *J. Chem. Phys.* **104** (1996) 742.
20. C. T. Rettner, *Phys. Rev. Lett.* **69** (1992) 383.
21. C. T. Rettner and D. J. Auerbach, *J. Chem. Phys.* **104** (1996) 2732.
22. M. Persson and B. Jackson, *J. Chem. Phys.* **102** (1995) 1078.
23. P. Kratzer, *J. Chem. Phys.* **106** (1997) 6752.
24. G. Eilmsteiner, W. Walkner, and A. Winkler, *Surf. Sci.* **352** (1996) 263.
25. Th. Kammler, J. Lee, and J. Küppers, *J. Chem. Phys.* **106** (1997) 7362.
26. S. Wehner and J. Küppers, *J. Chem. Phys.* **108** (1998) 3353.

27. Th. Kammler, S. Wehner, and J. Küppers, *J. Chem. Phys.* **109** (1998) 4071.
28. Th. Kammler, D. Kolovos-Velliantis, and J. Küppers, *Surf. Sci.* **460** (2000) 9.
29. G. Ertl, in: *Catalysis: Science and Technology* (eds. J. R. Anderson and M. Boudart), Vol. **4**, Springer, 1983, p. 209.
30. J. A. Dumesic, G. W. Huber, and M. Boudart, in: *Handbook of Heterogeneous Catalysis* (eds. G. Ertl, H. Knözinger, F. Schüth, and J.Weitkamp), Vol. **1**, Wiley, 2008, p. 1.
31. T. Stoltze and J. K. Nørskov, in: *Handbook of Heterogeneous Catalysis* (eds. G. Ertl, H. Knözinger, F. Schüth, and J. Weitkamp), Vol. **3**, Wiley, 2008, p. 1429.
32. V. P. Zhdanov, *Surf. Sci. Rep.* **45** (2002) 231.
33. P. Stoltze, *Prog. Surf. Sci.* **65** (2000) 65.
34. Y. Wang, K. Jacobi, W. D. Schoene, and G. Ertl, *J. Phys. Chem. B* **109** (2005) 7883.
35. S. T. Hatscher, Th. Fetzer, E. Wagner, and H. G. Kneuper, in: *Handbook of Heterogeneous Catalysis* (eds. G. Ertl, H. Knözinger, T. Schüth, and J. Weitkamp), Vol. **5**, Wiley, 2008, p. 2575.

CHAPTER 6

MECHANISMS OF HETEROGENEOUS CATALYSIS

The selective oxidation of ammonia, as discussed at the end of Chapter 5, was an example for the information to be obtained about catalytic reactions along the surface science approach. This chapter will now illustrate in some detail with three case studies how far the mechanisms of catalytic reactions can be elucidated and their kinetics described in the framework of the existing models.

6.1. SYNTHESIS OF AMMONIA ON IRON

The synthesis of ammonia from its elements, $N_2 + 3H_2 \rightarrow 2NH_3$, is one of the largest industrial processes based on heterogeneous catalysis [1–4]. Almost 90% of the present world capacity of about 200 million tons per year is used for the production of fertilizers. Almost all plants employ promoted catalysts based on iron, quite similar to the one developed by A. Mittasch in 1910 at the Badische Anilin and Sodafabrik (BASF) [5]. The above reaction could be successfully realized for the first time in the laboratory in 1909 by Fritz Haber, and was then transferred into a technical

---

Reactions at Solid Surfaces. By Gerhard Ertl
Copyright © 2009 John Wiley & Sons, Inc.

process by Carl Bosch at BASF, where the operation of first plant started in 1913.

Remarkably, despite this enormous technical importance and numerous laboratory studies, the actual mechanism of the Haber–Bosch process remained unclear for many decades. There was agreement that adsorption of nitrogen is the rate-limiting step, but the question whether the nitrogen species involved in the reaction is molecular or atomic in nature was not conclusively resolved [6].

The actual problem in studying the surface chemistry of the "real" catalyst becomes evident from the inspection of an electron micrograph from the Mittasch catalyst reproduced in Fig. 6.1 [7]. The "doubly promoted" catalyst is formed from a precursor consisting essentially of Fe_3O_4 with small concentrations of potassium, aluminum, and calcium oxides as listed in the first row of the table of Fig. 6.1. The surface composition differs considerably from that of the bulk and changes further upon reduction. The working catalyst consists of particles with about 30 nm size and a specific surface area of around $20\,m^2/g$. Under reaction conditions, these

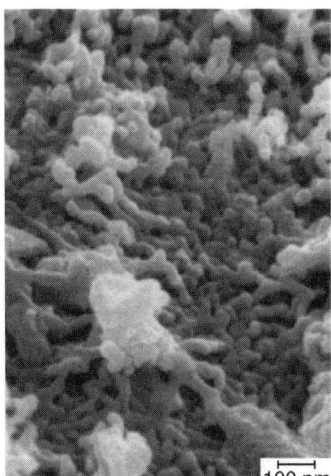

Technical conditions: $T \approx 400°C$, $p \approx 300$ bar
promoted iron catalyst

BASF S6-10 Catalyst (at. %)

	Fe	K	Al	Ca	O
Bulk composition	40.5	0.35	2.0	1.7	53.2
Surface					
Unreduced	8.6	36.2	10.7	4.7	40.0
Reduced	11.0	27.0	17.0	4.0	41.0
Cat. active spot	30.1	29.0	6.7	1.0	33.2

FIGURE 6.1. Surface topography and chemical composition of an industrial ammonia synthesis catalyst [7].

particles are reduced into metallic iron covered by a submonolayer of adsorbed K + O that acts as "electronic" promoter, as will be discussed below. The arrangement of active particles is stabilized against thermal sintering by a framework of Al_2O_3 (+CaO) that act as "structural" promoters. The active component itself will expose different crystal planes apart from various defects, and all these various structural parameters are presumably exhibiting varying reactivity.

The mechanism of the reaction could be clarified by means of the surface science approach. Earlier investigations with the "real" doubly promoted catalyst had revealed that the probability for adsorption (i.e., the sticking coefficient) is only of the order 10^{-6} and that this process is activated [9]. In Fig. 2.2, the variation of the relative surface concentration of adsorbed N atoms with exposure to gaseous N_2 at 693K on the Fe(1 1 0), (1 0 0), and (1 1 1) surfaces is shown. From these data initial sticking coefficients of 7×10^{-8}, 2×10^{-7}, and 4×10^{-6} were derived for the respective planes. The value for Fe(1 1 1) is of the same order of magnitude as that previously derived for the industrial catalyst [9]. Equally remarkable, the values for the sticking coefficients derived at very low pressures are quite similar to the reaction probabilities for different Fe single crystals determined at 20 bar pressure, as presented in Fig. 6.2 [10,11]. These results demonstrate that data obtained with single-crystal surfaces can be transferred across the "pressure gap" to high pressures and are also reflecting the behavior of "real" catalysts.

Both the most active Fe(1 1 1) and (2 1 0) surfaces are characterized by the existence of the so-called C7 sites. These sites had already been suggested to exhibit the highest activity on the basis of investigations by Mößbauer spectroscopy [12]. The activities of the Fe(1 0 0) and (1 1 0) surfaces were, on the other hand, markedly increased if these were pretreated by ammonia at 723K [13]. This procedure is associated with a pronounced restructuring of the surface that thereby becomes more "(1 1 1)-like," so the

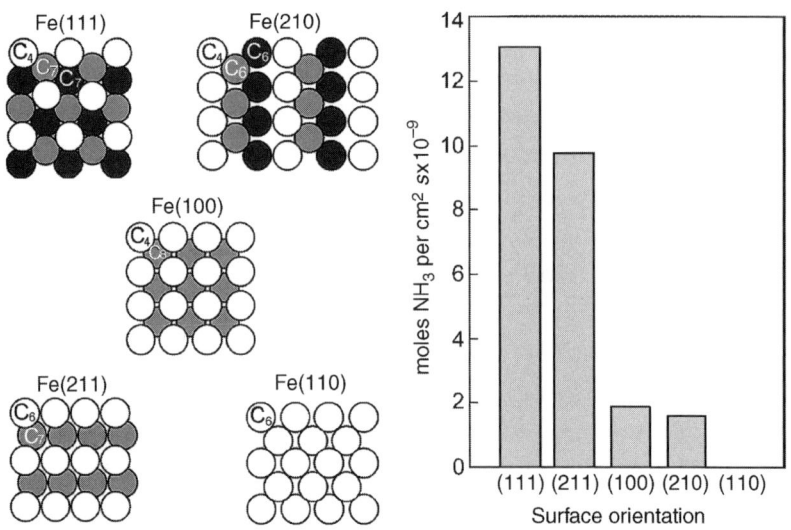

FIGURE 6.2. The specific reactivity for ammonia synthesis of various Fe single-crystal surfaces at 20 atm and 798K [10].

Fe(1 1 1) surface can indeed be regarded as representing the nature of the active surface under working conditions.

Dissociative adsorption of nitrogen proceeds through molecularly adsorbed intermediate states, where the behavior of the Fe(1 1 1) surface is of particular interest: Exposure to N_2 at low temperature ($<$ 80K) causes the formation of an adsorbed species (α-state) with an adsorption energy of 24 kJ/mol and a N–N stretch frequency of 2100 cm^{-1} (for $^{15}N_2$), which is very close to the value of the free N_2 molecule. Together with the analysis of data from photoelectron spectroscopy [14,15], this led to the conclusion that this is a physisorbed species with its molecular axis perpendicular to the surface.

A second more tightly held molecular state is formed at slightly higher temperature with a sticking coefficient of about 10^{-2} and an adsorption energy of 31 kJ/mol (α-state) [16]. Interestingly, the frequency of the N–N stretch vibration is appreciably lowered to 1490 cm^{-1} ($^{15}N_2$), indicating a considerable reduction of the N–N bond order [14,17]. Comparison with coordination compounds led to the conclusion that now the N_2 molecule is bonded "side on" with one of the N atoms attached to the C7 site. The lowering of the N–N stretch frequency reflects a weakening of the N–N bond, and an increase of this length is characteristic of a π-bonding mechanism in which the antibonding π-orbitals of N_2 act as acceptors for electrons from the metal substrate.

Warming up the Fe(1 1 1) surface covered by α-N_2 leads to dissociation as well as desorption, as reflected by the disappearance of the N–N stretch vibration and the appearance of a new band at 450 cm^{-1}, which is attributed to the metal–N vibration (β-state). This bond is rather strong, and recombinative desorption takes place only around 700K. This process is usually associated with a pronounced reconstruction of the surface, and the resulting structures are denoted as "surface nitrides." An exception is found with the Fe(1 0 0) surface where a simple c2 × 2 structure is formed whose structure was analyzed by LEED and is reproduced in Fig. 6.3 [18]. The N atoms occupy fourfold hollow sites with their plane only about 0.03 nm above that formed by the topmost Fe atoms. This structure is very similar to that of the (0 0 2) plane of bulk Fe_4N, and hence termed as "surface nitride." (It should be noted that bulk iron nitrides will not be stable under the conditions of ammonia synthesis for thermodynamic reasons, but may be readily formed in NH_3/H_2 mixtures providing high "virtual" N_2 pressures.)

If we return to the Fe(1 1 1) surface, the progress of dissociative nitrogen adsorption can now be rationalized in terms of the schematic potential diagram depicted in Fig. 6.4. Detailed analysis of the kinetics at zero coverage revealed for the sticking coefficient in the α-state a value of 10^{-2}, and from the temperature dependence of the formation of the atomic β-state in terms of the Arrhenius parameters, $s_\beta = v_0 \exp(-E^*/kT)$, with $v_0 = 2.2 \times 10^{-6}$ and $E^* = -3$ kJ/mol (i.e., a slight decrease of the overall sticking coefficient with temperature) [16]. These data are plotted in Fig. 6.5 as open circles together with those from another low pressure

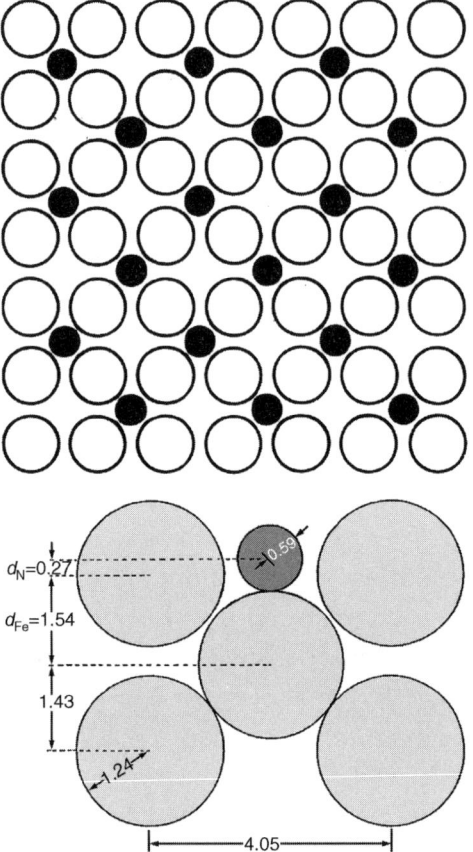

FIGURE 6.3. Structure of the c2 × 2 phase formed by adsorbed N atoms on a Fe(1 0 0) surface (distances in Å) [18].

study (open squares) [19] and from another one with N_2 pressures up to 600 mbar (filled squares) [20]. They provide a consistent picture for the precursor-mediated (indirect) dissociative nitrogen adsorption that also underlies the microkinetic modeling of the overall reaction to be discussed below.

An apparently contradicting model resulted from molecular beam studies [20]. For low kinetic energies, the sticking coefficient was again found to be around 10^{-6}, but an increase by more than four orders of magnitude with increasing translational energy of

Synthesis of Ammonia on Iron

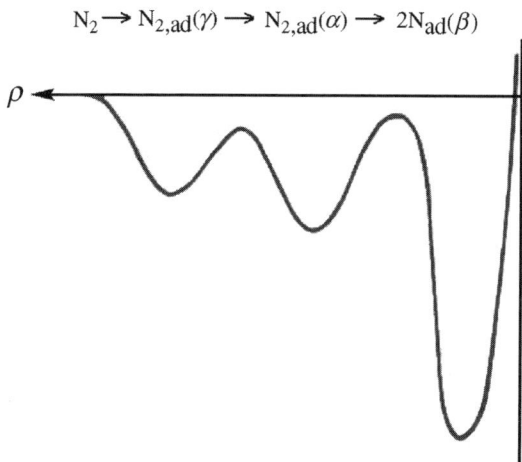

FIGURE 6.4. Schematic potential diagram for the dissociative adsorption of nitrogen on a Fe(1 1 1) surface.

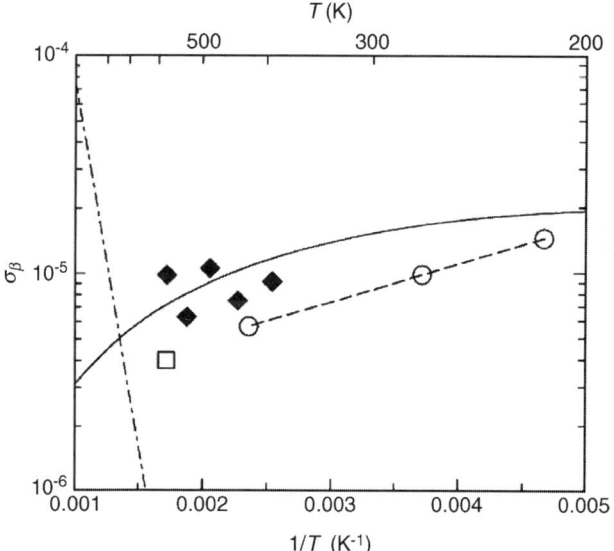

FIGURE 6.5. The sticking coefficient for dissociative nitrogen adsorption on Fe (1 1 1), open circles [16], open square [19], and filled squares [20], experimental data. Full line and desk-dotted line theoretical data for the indirect and direct path, respectively [22].

the incident N_2 molecules suggests crossing of an activation barrier by direct collision from the gas phase.

This problem could be solved in a theoretical study. First of all, analysis of the kinetics in terms of transition-state theory reveals that the Arrhenius parameters are temperature dependent, and as a consequence the *true* activation energy is $+3\,\text{kJ/mol}$, that is, slightly positive. Thus, all the experimental data of Fig. 6.5 can be consistently modeled (i.e., full line) without any further assumptions. Along these lines the very small preexponential factor of the experimental sticking coefficient can be traced back to a large entropic barrier between the free N_2 molecule and its transition state (TS). The "direct" path as suggested by the molecular beam experiments is, on the other hand, associated with a high activation energy that explains the strong increase of the sticking coefficient with kinetic energy. The theoretical data resulting for this channel are plotted as dash-dotted line in Fig. 6.5, indicating that under normal thermal conditions the "indirect" path will be dominating.

The sticking coefficient for dissociative nitrogen adsorption will be markedly increased by the presence of preadsorbed potassium up to about 4×10^{-5} for Fe(1 1 1) [23] as well as with other iron surfaces [24]. On Fe(1 1 1), the adsorption energy for α-N_2 is locally increased from 31 to $44\,\text{kJ/mol}$, giving rise to two peaks (α_1 and α_2) in the thermal desorption spectra. Simultaneously, the activation energy for dissociation is lowered. Interpretation of this effect is achieved by an electrostatic model [24,25]. As reflected by the strong decrease of the work function, potassium adsorption is associated with a pronounced transfer of electronic charge to the substrate. An N_2 molecule will then experience a stronger "backbonding" effect from the metal to its π^* orbitals, where the N–N bond is weakened and the M–N_2 bond is strengthened, as illustrated in Fig. 6.6a. The surface of the actual catalyst is not covered by K alone (which would thermally desorb under reaction conditions), but by a mixed K + O adlayer

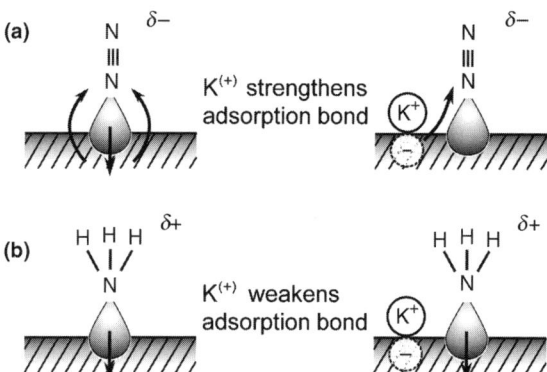

FIGURE 6.6. Cartoon illustrating the effect of a K promoter on the bonding of N_2 (a) and NH_3 (b).

with about 1:1 composition. Both components mutually stabilize each other against thermal desorption and reduction, respectively. Since O is electronegative, the promoter effect of K alone will be partly reduced, but the sticking coefficient is still enhanced [26].

There is an additional effect of the potassium promoter to be discussed. It was found that its effect is more pronounced at higher reaction rates, that is, higher stationary concentrations of adsorbed ammonia. Adsorbed NH_3 blocks sites for the adsorption of the reactants. Its dipole moment has, however, the opposite sign of that of α-N_2, so the presence of K will lower its adsorption energy and hence reduce its site-blocking effect [27], as illustrated in Fig. 6.6b.

The other surface species involved in the reaction will only be briefly discussed.

Hydrogen is dissociatively adsorbed with high sticking coefficient with adsorption energies around 100 kJ/mol [28], which is not significantly affected by the presence of potassium. At the high temperatures of ammonia synthesis, the lifetime of adsorbed H atoms will be very short, and they may be considered as a highly mobile two-dimensional gas.

Binding of adsorbed NH_3 to the surface occurs through the low-electron pair of the N atom, as sketched in Fig. 6.6b, with high sticking coefficient and an adsorption energy of about 70 kJ/mol [29,30]. Decomposition of adsorbed NH_3 has to proceed through the same steps as its formation, so the identification of these intermediates at low pressures will identify the last steps in the synthesis reaction.

In fact, heating up an ammonia-covered iron surface leads not only to desorption but also to stepwise dissociation, where the intermediate formation of $NH_{2,ad}$ [29] and NH_{ad} [31] could be demonstrated.

If all the experimental experience is put together, the mechanism of this reaction can be formulated and its potential energy diagram established, as represented in Fig. 6.7 [8]. Homogeneous reaction in the gas phase would be energetically prohibited because of the large dissociation energies of the first steps. The alternate reaction pathway through the catalyst overcomes this problem, since the energy gain associated with dissociative adsorption overcompensates these dissociation energies, and the first steps become even exothermic. Dissociative nitrogen adsorption is nevertheless rate limiting, not because of a high activation barrier, but rather because of the low preexponential factor.

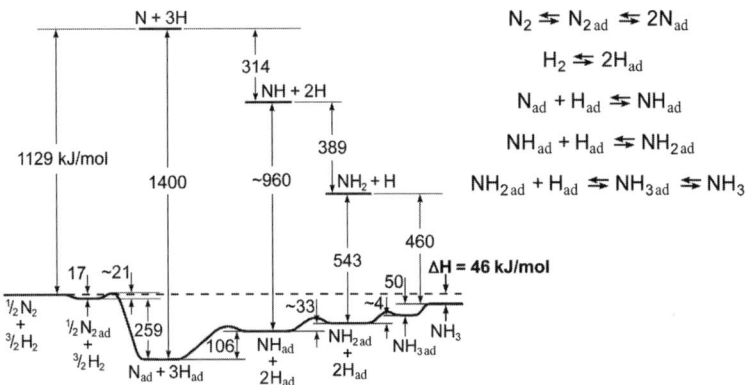

FIGURE 6.7. Mechanism and potential energy diagram for ammonia synthesis on iron (energies in kJ/mol) [8].

The subsequent hydrogenation steps are energetically uphill, but the necessary energies are readily provided by the reaction temperature of about 700K.

Under the assumption that dissociative nitrogen adsorption is rate limiting and the heat of adsorption varies linearly with coverage, Temkin and Pyzhev [32] had derived rate equations that had been extensively applied to model the kinetics under "real" conditions. However, it was pointed out by Boudart [33] that integration of the rates over (assumed) nonuniform adsorption energetics yields an expression for the overall turnover frequency that is interestingly similar to that resulting for a uniform surface with Langmuir kinetics for which the parameters at low coverages are taken. It was, therefore, concluded [34] that modeling of the kinetics with a "real" catalyst based on data from the surface science approach might indeed be successful.

Such microkinetic modeling was first performed by Stoltze and Nørskov [35,36] who assumed that dissociative nitrogen adsorption is rate limiting and all other steps are in equilibrium through Langmuir-type isotherms. The experimental data for the K-promoted Fe(1 1 1) surface were used for the rate of dissociative nitrogen adsorption. Comparison of the resulting ammonia yields with those determined in industrial plants with a commercial Topsoe KM1 catalyst up to pressures of 300 atm (i.e., more than nine orders of magnitude higher than with the surface science studies!) yielded remarkably good agreement. Figure 6.8 shows how this agreement (better than within a factor of 2) extends over a wide range of conditions [35].

Similar microkinetic models developed by other groups [37–39] revealed similar close agreement between the experimental data from single crystal studies and "real" catalysis, and at a symposium in honor of H. Topsoe and A. Nielsen, two of the leaders in industrial ammonia synthesis, general agreement was reached that the main aspects of catalytic ammonia synthesis are now essentially understood [40].

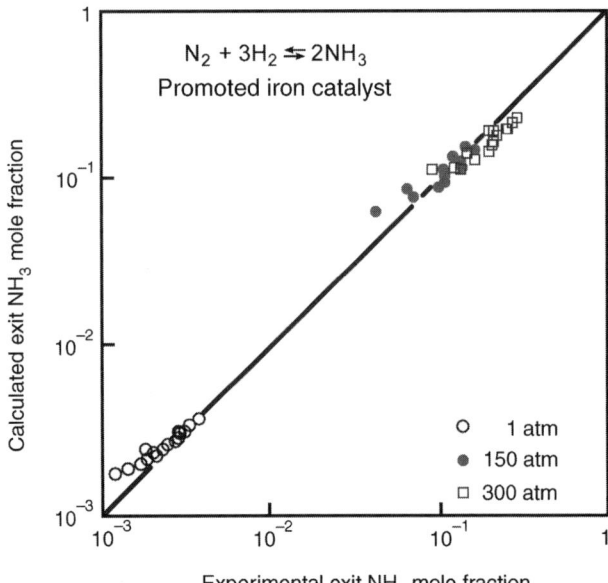

FIGURE 6.8. Microkinetics of ammonia synthesis on iron catalysts: comparison of the yields calculated on the basis of the mechanism presented in Fig. 6.7 (y-axis) with experimental data from industrial plants (x-axis). The straight line marks perfect agreement [35].

6.2. Synthesis of Ammonia on Ruthenium

In the original work on catalytic ammonia synthesis, Haber [41] had used an osmium catalyst, but this metal was much too expensive to be the basis of the large-scale industrial plants. In the long search for alternatives to the Mittasch catalyst, alkali-promoted ruthenium was found to exhibit specific activity, which is even superior to the iron catalyst [42] and which was subsequently developed to an industrial catalyst [43]. The Mittasch catalyst is cheap and the alumina promoter provides a high specific surface area. This situation is different with Ru catalysts that are prepared as small particles on a suitable support. Figure 1.1 was a typical electron microscopic picture from such a catalyst particle on $MgAl_2O_4$ (spinel) support [44].

In this case, the influence of the alkali promoter is even more pronounced than with Fe, and also the nature of the support exerts a marked influence. This becomes evident from Fig. 6.9 in which the steady-state yields of Cs-promoted and unpromoted Ru catalysts on Al_2O_3 and MgO supports at atmospheric pressure are plotted [45].

The reaction proceeds along the same elementary steps as with the Fe catalyst, and again successful microkinetic modeling on the basis of experimentally derived parameters could be achieved [46]. Again, dissociative nitrogen adsorption is rate limiting, where the sticking coefficient is markedly affected by the presence of atomic steps [47] whose role as "active sites" had been discussed in Chapter 5.

On the Ru(0001) surface, the N_2 molecule is terminally bonded "on top" with its molecular axis perpendicular to the surface [48]. After dissociation, the N atoms are located in threefold coordinated sites and form two ordered phases with coverages $\Theta_N = 0.25$ and 0.33, respectively, whose structures as

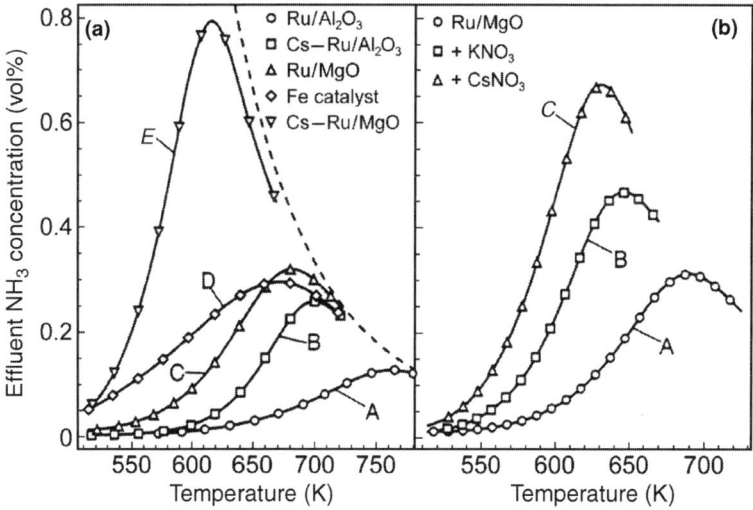

FIGURE 6.9. Steady-state ammonia yields of various Ru catalysts as a function of temperature at atmospheric pressure [45].

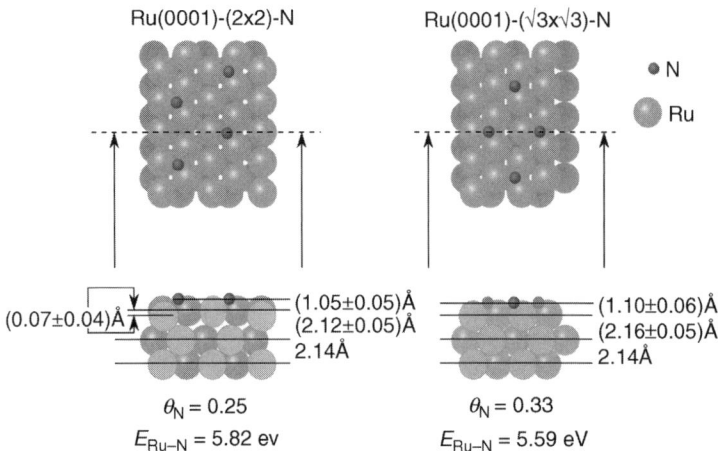

FIGURE 6.10. Ordered phases formed by adsorbed N atoms on a Ru(0001) surface at coverages $\theta = 0.25$ and 0.33 [49].

determined by LEED are depicted in Fig. 6.10 [49]. The energetics of dissociative adsorption has been treated theoretically with inclusion of the action of Na and Cs promoters and the resulting potential diagram is reproduced in Fig. 6.11 [50]. These

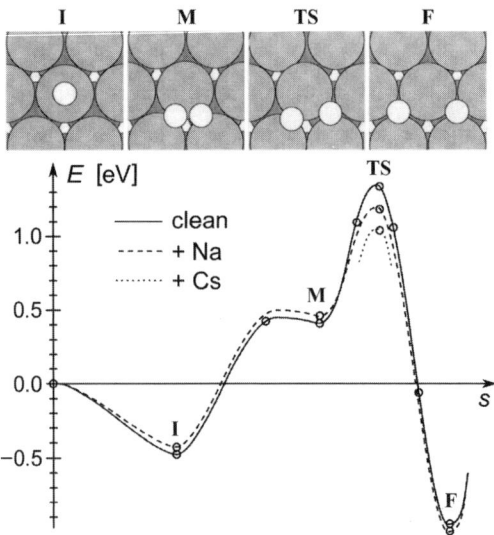

FIGURE 6.11. Theoretical potential energy diagram for dissociative nitrogen adsorption on bare and alkali-promoted Ru(0001) surfaces [50].

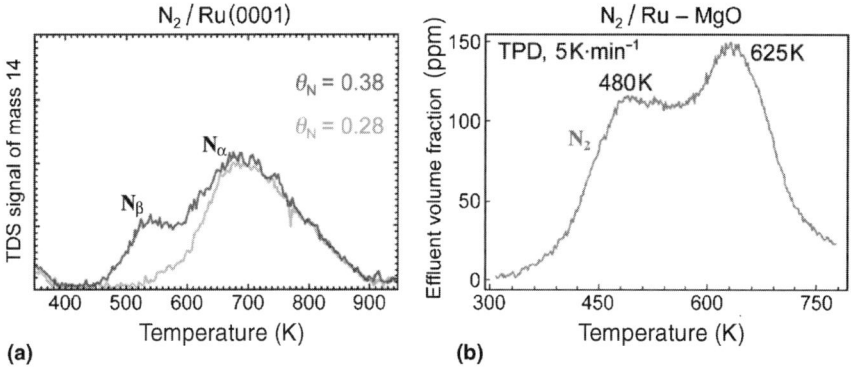

FIGURE 6.12. Thermal desorption spectroscopy of N_2 from Ru surfaces. (a) Ru (0 0 0 1) single-crystal surface [52]. (b) Ru catalyst supported on MgO [53].

data clearly show the stabilization of the transition state by the presence of the alkali promoter. Adsorbed nitrogen atoms repel each other, so the higher coverage phase at $\Theta_N = 0.33$ has a lower adsorption energy than that at $\Theta_N = 0.25$. This difference shows up in thermal desorption spectroscopy (TDS) as two peaks, as reproduced in Fig. 6.12a [52]. Analogous experiments with a MgO-supported Ru catalyst at atmospheric pressure again reveal two peaks in the TDS data (Fig. 6.12b) [53], demonstrating the similarity between the "real" catalysts and the single-crystal surface. (The peak positions in Fig. 6.12 differ slightly from each other because of the different heating rates.)

As a next step, DFT calculations were employed to evaluate the potential energy diagram for the whole reaction sequence over (unpromoted) flat and stepped Ru(0 0 0 1) surfaces [51]. The result is reproduced in Fig. 6.13 and demonstrates that the step sites are much more reactive in dissociating N_2 than terrace sites in full agreement with experiment.

The final step toward a complete first-principles description of the Ru-catalyzed ammonia synthesis was performed by including the full complexity of interactions and reaction paths and comparing the resulting rates with experimental data from supported Ru particles, such as reproduced in Fig. 1.1 [44]. The only information

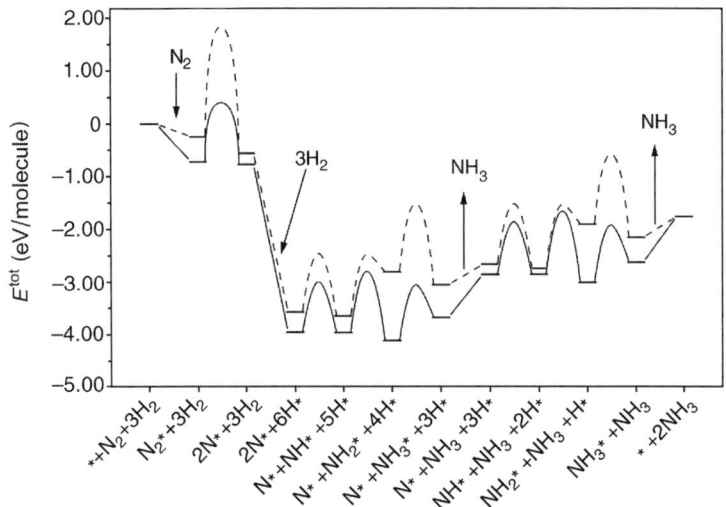

FIGURE 6.13. Theoretical potential energy diagram for NH_3 synthesis on stepped (full line) and flat (broken line) Ru(0 0 0 1) surfaces [51].

needed from the actual catalyst was the density of active sites that was derived from the particle size distribution. Comparison of the measured and calculated NH_3 yields for various conditions is shown in Fig. 6.14. If one takes into consideration that this result

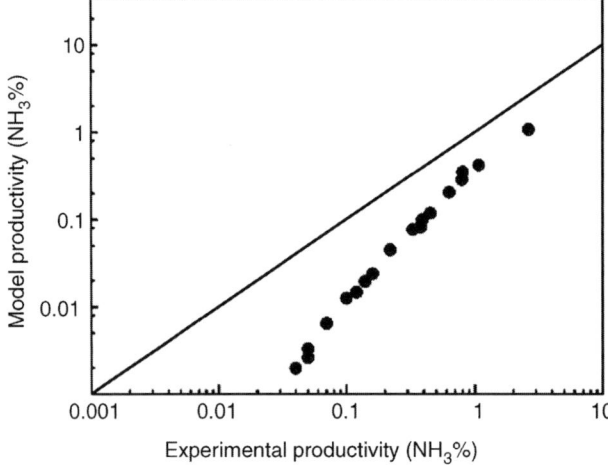

FIGURE 6.14. Comparison of experimental and theoretical ammonia yields on supported Ru catalysts (data points) [44]. The full line would represent perfect agreement.

was obtained without any fit parameters, the agreement is really extraordinary, in particular if the uncertainty inherent in the DFT method is taken into account. This example convincingly demonstrates how far advanced theoretical understanding of heterogeneous catalysis may become.

6.3. OXIDATION OF CARBON MONOXIDE

The oxidation of carbon monoxide takes place in the car exhaust cleaning by means of catalysts based on the platinum metals and represents the simplest heterogeneously catalyzed reaction [54,55]. It involves chemisorption of CO and dissociative chemisorption of oxygen, and $CO_{ad} + O_{ad}$ react with each other to CO_2 via the Langmuir–Hinshelwood mechanism. As an example, Fig. 6.15 shows the ordered structures formed by these adsorbates on a Rh(1 1 1) surface [56]. The CO molecules are in this case bonded to the surface "on top" and always exhibit the tendency to form densely packed adlayers if the coverage becomes high enough (Fig 6.15a). The O atoms, on the other hand, occupy

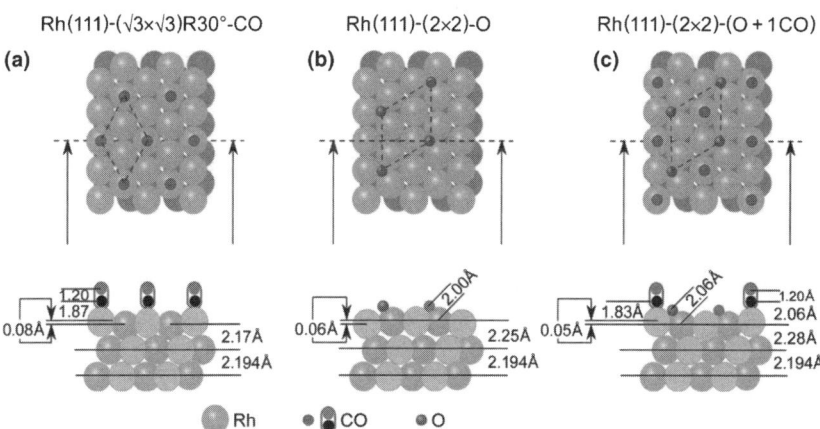

FIGURE 6.15. Ordered structures formed on a Rh(1 1 1) surface [56]. (a) $\sqrt{3}$ phase of CO. (b) 2×2 phase by O. (c) 2×2 phase by coadsorption of O + CO. (See color insert.)

threefold coordinated sites and form an open mesh of a 2×2 structure (Fig. 6.15b). Since for the dissociation of O_2 an ensemble of neighboring empty surface atoms is required, this process will become inhibited as soon as the CO coverage exceeds a critical value. The open structure of the O adlayer, on the other hand, still permits adsorption of CO, causing formation of a mixed adlayer (Fig. 6.15c), where the two reactants are in close contact. Under steady-state flow conditions, this asymmetric behavior of the two adsorbates leads soon to full coverage of the surface by adsorbed CO that prevents oxygen adsorption and inhibits the reaction. This problem can only be overcome if the temperature is high enough (>450K) to enable continuous desorption of part of the adsorbed CO so that O_2 from the gas phase can compete for the free adsorption sites. This is the reason why the car exhaust catalyst operates only at elevated temperatures.

In the following, we will concentrate on the behavior of the Pt(1 1 1) surface. In contrast to Pt(1 0 0) and Pt(1 1 0) (which will be discussed in Chapter 7), this most densely packed plane does not reconstruct.

The progress of the reaction is sketched in Fig. 6.16. In this case, adsorbed CO forms a densely packed $c2 \times 4$ layer occupying both the "on top" and "bridge" sites [57], while adsorbed O gives rise to the same 2×2 structure as with Rh(1 1 1). Consequently, the same mixed structure, as depicted in Fig. 6.15c, will be formed if both adsorbates are present on the surface. The operation of the Langmuir–Hinshelwood mechanism was established through modulated molecular beam experiments that enabled derivation of the kinetic parameters in the framework of the Langmuir approximation [58]. (The CO_2 formed is so weakly held at the surface that it is immediately released into the gas phase [59].)

There are, however, strong indications that the actual mechanism involves more complex aspects. The activation energy of 100 kJ/mol indicated in Fig. 6.16 holds only for small coverages, while it is only half as high at higher coverages [58]. From

OXIDATION OF CARBON MONOXIDE 141

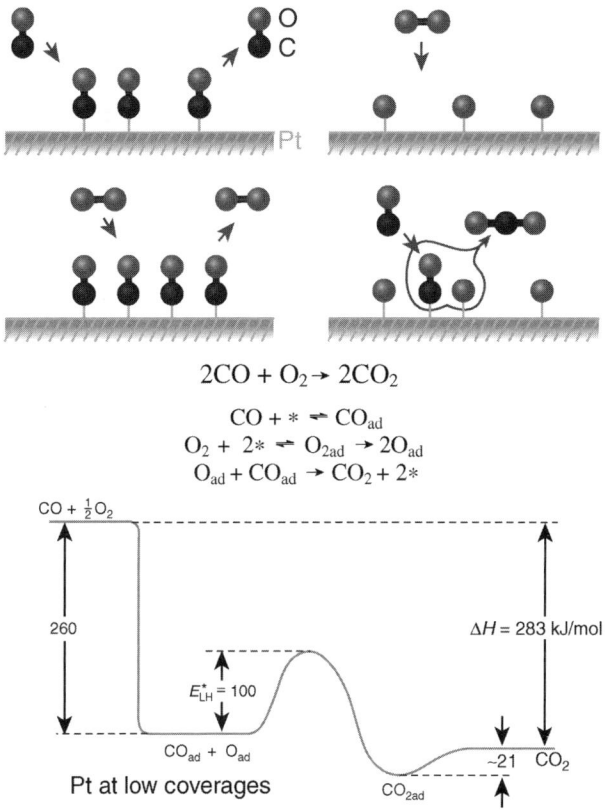

FIGURE 6.16. Mechanism and potential diagram for catalytic CO oxidation at a Pt(1 1 1) surface at low coverages (energies in kJ/mol). (See color insert.)

temperature-programmed desorption studies, it had previously been concluded that the reacting O and CO species are not randomly distributed, but the reaction takes place at the perimeters of islands [61].

To elucidate the mechanism *on atomic scale*, the reaction was investigated by means of scanning tunneling microscopy (STM) [62]. The experiments concentrated on the LH step $O_{ad} + CO_{ad} \rightarrow CO_2$, where an O-covered surface was exposed to CO and the change of the surface structure was continuously monitored. A series of STM images is reproduced in Fig. 6.17. After

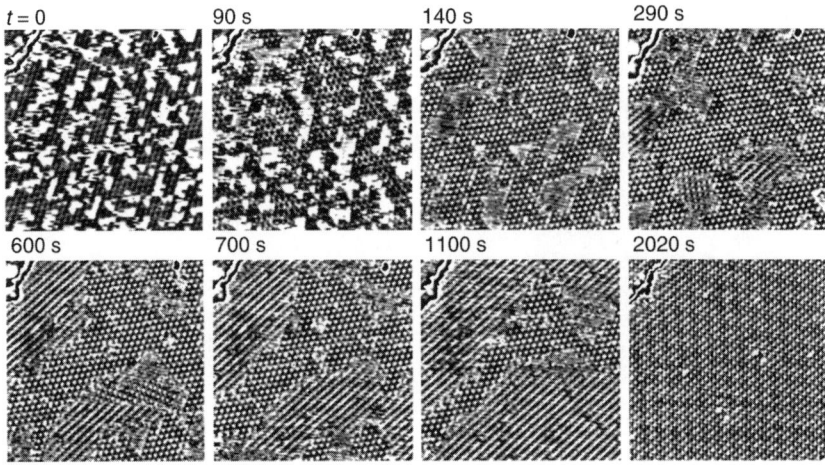

FIGURE 6.17. Series of STM images from a Pt(1 1 1) surface initially covered by adsorbed O atoms ($t=0$) and then exposed to 5×10^{-8} mbar CO at 247K. Image size 18×17 nm^2 [62].

preparation of a submonolayer of O_{ad} ($t=0$) at $T=247$K, CO was continuously supplied at $p_{CO} = 5 \times 10^{-3}$ mbar. This partial pressure offers about 1 monolayer of CO in 100 s. At first coadsorption of CO causes some better ordering of O_{ad} in 2×2 domains ($t=90$ and 140 s) of the mixed phase, while oxygen-free regions in between these patches are occupied by (mobile) CO. After 290 s, the 2×2 fraction of the surface had started to shrink because of the progress of the reaction. Simultaneously, new patches of the c4 $\times$ 2 structure were formed by densely packed CO [63] until the reaction is completed after 2020 s and the surface fully CO covered. Evidently, the reaction proceeds preferentially not within the mixed phase where both reactants are in intimate contact as in Fig. 6.15c, but at the boundaries from the adjacent CO overlayer. As depicted in Fig. 6.18, for CO molecules on bridge sites, the situation for reaction with an adjacent O atom is obviously energetically more favorable than within the unit cell of the mixed 2×2 phase. This conclusion is supported by the results of DFT calculations, as illustrated in Fig. 6.19 [60]: If a CO molecule at on top position approaches an O atom on a threefold site (as

OXIDATION OF CARBON MONOXIDE 143

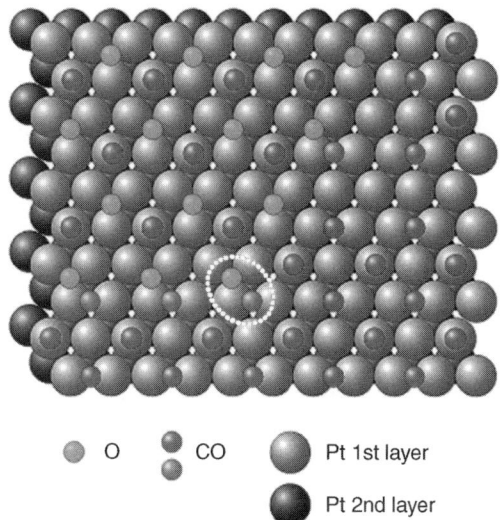

FIGURE 6.18. Cartoon illustrating the boundary between the mixed 2 × 2 (O + CO) phase and the c2 × 4 phase of CO on Pt(1 1 1). (See color insert.)

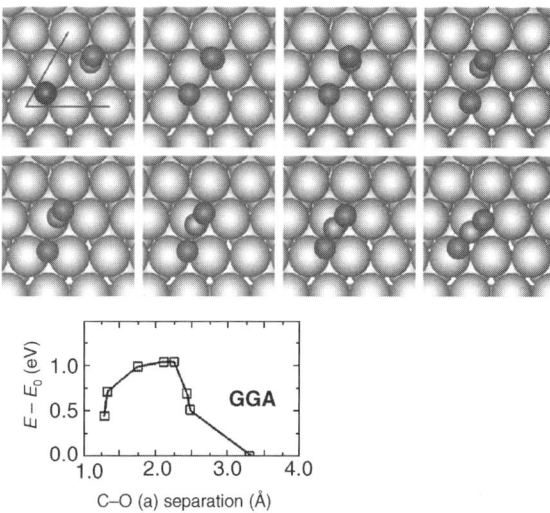

FIGURE 6.19. DFT calculations for the energy on a Pt(1 1 1) surface if O_{ad} and CO_{ad} approach each other to form CO_2 [60]. (See color insert.)

within the unit cell of the mixed phase), an activation energy of 1.0 eV (i.e., 100 kJ/mol) has to be surmounted for CO_2 formation. This is exactly the value determined experimentally for low coverages as outlined above, where the reactants approach each other from longer distances. Figure 6.19 reveals that this activation energy is only half as high if the CO molecule has already occupied a bridge site next to the O atom. This is just the experimental value at high coverages and represents the situation depicted in Fig. 6.18. Thus, it becomes plausible why the reaction proceeds preferentially at the boundaries and not within the uniform mixed phase.

The STM data were even subject of quantitative analysis of the kinetics. The rate was evaluated by determining the 2×2 covered part of the surface ($\hat{=}$ O coverage Θ_O) as a function of the time at constant p_{CO}. If normalized to the length of the domain boundaries L between CO and O domains, this rate practically becomes constant (squares in Fig. 6.20). The microscopic (i.e., true) rate can, therefore, be expressed as $r' = k' \cdot L$, where the CO coverage was always at saturation as checked by varying the CO pressure over one order of magnitude. Measurements at varying temperature yielded $E^* = 0.49$ eV and $\nu = 4 \times 10^{21}$ particles cm^{-2} s^{-1} for the kinetic parameters. These data can be compared with macroscopic parameters derived from molecular beam experiments

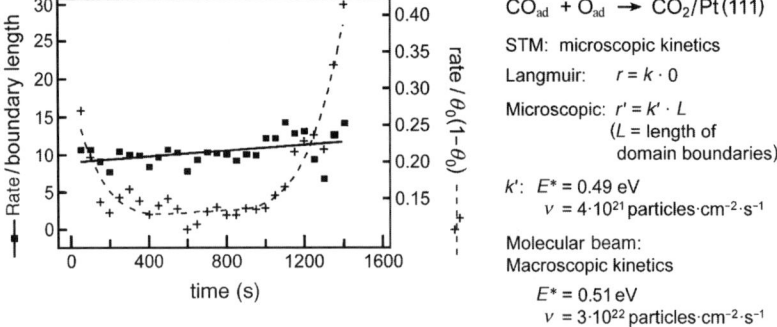

FIGURE 6.20. Kinetics of the reaction $O_{ad} + CO_{ad} \rightarrow CO_2$ on a Pt(1 1 1) surface at 247K, as derived from the STM data shown in Fig. 6.17 [62].

within the LH (i.e., mean field) approximation under comparable coverage conditions. There values of $E^* = 0.51$ eV and $\nu = 3 \times 10^{22}$ particles cm^{-2}s^{-1} were derived that are in excellent agreement. The question arises why is the LH approach so successful in this case if the conditions (random distribution of the reactants and spatially uniform reactivity) are definitely not fulfilled? In this case, the rate is expressed as $r = k \cdot \Theta_O \cdot \Theta_{CO}$, or if we set $\Theta_O + \Theta_{CO} = 1$, then $r = k\Theta_O(1-\Theta_O)$. If we determine Θ_O by counting the number of O atoms from the fraction of the surface occupied by 2×2 patches, the normalized rate as marked by crosses in Fig. 6.20 results. It turns out that the LH *rate constant* k is indeed constant over a fairly wide range of coverages. This result is in fact striking, but can be considered as justification of the mean field approximation in microkinetic analysis even in cases where the basic assumptions are definitely not fulfilled.

The mechanism and kinetics of CO oxidation are quite similar for Pt, Pd, and Rh for single crystals and for supported catalysts at atmospheric pressure or under UHV conditions [54,55]. There is, however, a serious problem with Ru suggesting the existence of a pronounced "pressure gap": With supported catalysts at atmospheric pressure, Ru was found to be much more reactive than Rh, Pt, and Pd [64], while UHV studies with single-crystal surfaces found just the opposite in that Ru was practically inactive [54,65,66]. That this difference is not a structural effect was demonstrated by Peden and Goodman [67] who found that a Ru(0 0 0 1) surface again became very active when the reaction was performed near atmospheric pressure. This apparent "pressure gap" is however a "materials gap": It was found that in an oxidizing atmospheric pressure and at elevated temperature, the clean Ru(0 0 0 1) surface transforms slowly into a thin overlayer of RuO$_2$ exposing its (1 1 0) surface [68]. This transformation is illustrated in the STM image of Fig. 2.26a, while Fig. 6.21a shows a ball model of the resulting RuO$_2$(1 1 0) surface. The stoichiometric surface is terminated by O atoms, thereby bridging saturated

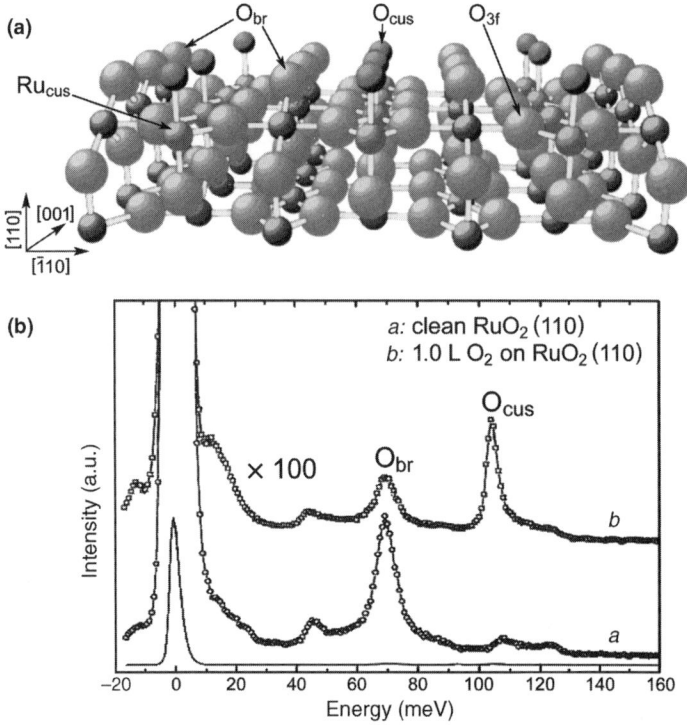

FIGURE 6.21. The $RuO_2(1\,1\,0)$ surface [69]. (a) Ball model with additional O atoms adsorbed on cus sites. (b) Corresponding vibrational spectrum. (See color insert.)

Ru atoms ($=O_{br}$), and by coordinatively unsaturated (cus) Ru atoms onto which dissociative adsorption of further oxygen ($=O_{cus}$) as well as of CO may take place.

These two surface oxygen species can be distinguished on the basis of their vibrational spectra, as reproduced in Fig. 6.21b [69]. This technique permits to follow the reactive interaction between O_{ad} and CO_{ad}, as illustrated in Fig. 6.22. The left part shows how an oxygen-saturated surface changes upon exposure to gaseous CO. The O_{ad}-derived bands are continuously replaced by those characteristic of adsorbed CO. On the right-hand site, on the other hand, the replacement of adsorbed CO takes place by exposure to O_2. Both processes occur readily with low activation energy

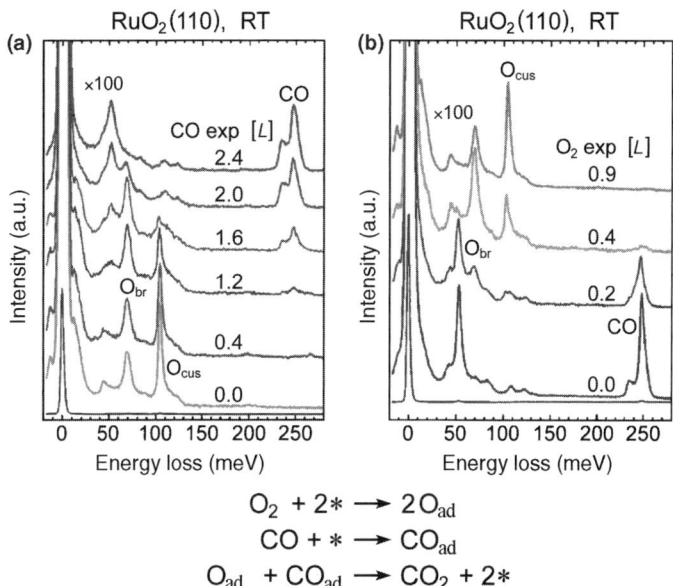

FIGURE 6.22. Vibrational spectra from a $RuO_2(110)$ surface showing the progress of the reaction between adsorbed O and CO. (See color insert.)

already at 300 K. Obviously, this system represents a clear example for the LH mechanism in that the same active sites (the Ru cus atoms) may be occupied by both reactants that then form the product by mutual interaction on neighboring sites. (It should, however, be mentioned that the O_{br} species are of comparable reactivity.) This situation has been modeled theoretically, and the resulting energy profile is reproduced in Fig. 6.23 [70]. By moving both O_{cus} and CO_{cus} together along the [0 0 1] direction, CO_2 is formed with an activation energy of 0.89 eV.

The steady-state kinetics is accordingly expected to exhibit the characteristics of an LH mechanism. This is qualitatively confirmed by the experimental data for the rate of CO_2 formation (expressed in terms of turnover frequency) as a function of p_{CO} at constant p_{CO_2} and T [71], as shown in Fig. 6.24. For comparison, this figure also represents the data resulting from the theory for the same set of parameters [72]. In this case, full *ab initio* calculations

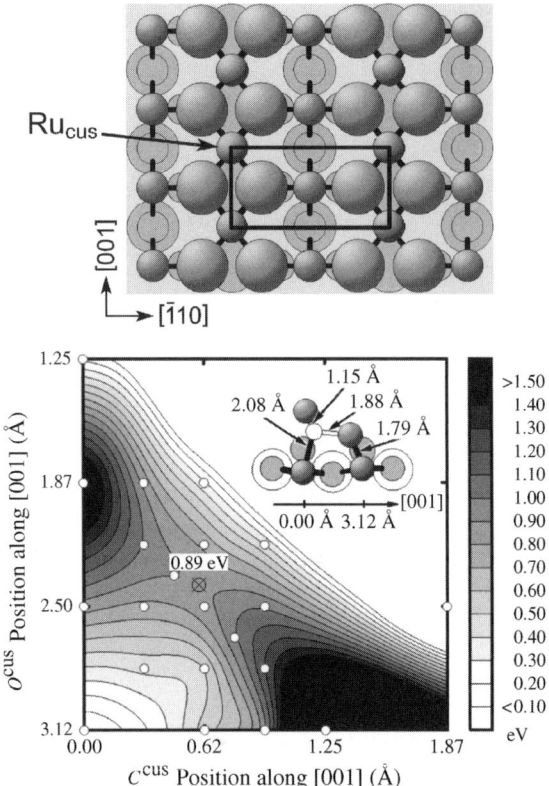

FIGURE 6.23. Theoretical energy diagram for the interaction between O and CO on a RuO$_2$(1 1 0) surface [70]. (See color insert.)

were made by combining DFT and statistical mechanics. The agreement between theory and experiment is remarkably good if one takes into consideration that the kinetics is expressed quantitatively in terms of turnover frequency and the calculations contain no adjustable parameters.

It was found that with this system the behavior of single-crystal surfaces (after transformation into RuO$_2$) under UHV conditions is identical to that of supported Ru catalysts at elevated pressures, so both the pressure and the materials gap are successfully bridged [73]. With supported catalysts, the most stable and

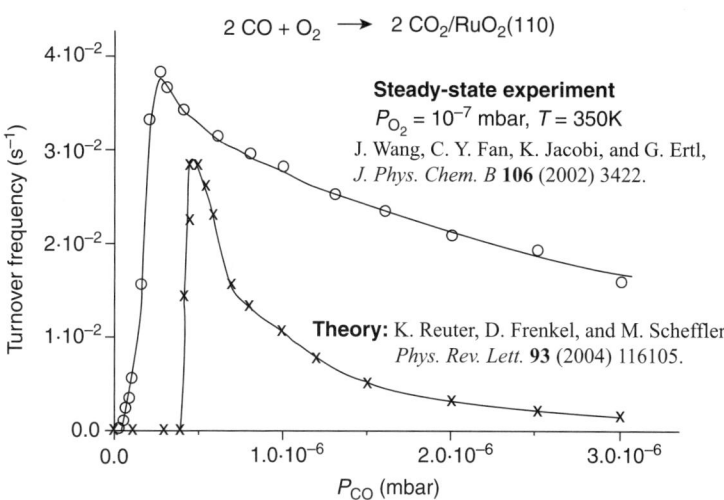

FIGURE 6.24. Steady-state rate of CO_2 formation in the catalytic oxidation of CO on a $RuO_2(110)$ surface as a function of p_{CO} at constant $p_{O_2} = 10^{-7}$ mbar and $T = 350$K: Experiment [71] and theory [72].

active state was identified with an ultrathin RuO_2 shell exposing the (1 1 0) and (1 0 0) surfaces coating a metallic Ru core, so the similarity with the single-crystal studies becomes evident [74].

6.4. Oxidation of Hydrogen on Platinum

The oxidation of hydrogen catalyzed by finely divided platinum was first observed by Döbereiner in 1823 [75] and prompted Berzelius to introduce the term "catalysis" [76]. One might expect that in this case the mechanism is comparably simple as with carbon monoxide oxidation. Both reactants are known to first adsorb dissociatively. While O_{ad} forms the known 2×2 structure presented several times before, with adsorbed H a problem arises: The maximum coverage was found to be one, but no ordered structures could be identified with LEED nor could the adsorbates be identified by STM. This is caused by the fact that this surface is energetically very smooth, and the first vibrational excitations need only a few kJ/mol [77–80]. As a consequence, at not extremely

$$2H_2 + O_2 \rightarrow 2H_2O \,/\, Pt$$

Mechanism (?):
- (1) $\quad O_2 + * \rightarrow 2O_{ad}$
- (2) $\quad H_2 + * \rightarrow 2H_{ad}$
- (3) $\quad O_{ad} + H_{ad} \rightarrow OH_{ad}$
- (4) $\quad OH_{ad} + H_{ad} \rightarrow H_2O_{ad}$
- (5) $\quad H_2O_{ad} \rightarrow H_2O_g + * \qquad (T \geq 170K)$

But

	$T < 170K$	$T > 230K$
	Induction period	No induction period
E^*	$\sim 12\,kJ/mol$	$> 150\,kJ/mol$

Important
- (6) $\quad H_2O_{ad} + O_{ad} \rightarrow 2OH_{ad}$

FIGURE 6.25. The mechanism of the $H_2 + O_2$ reaction on Pt(1 1 1) [89].

low temperatures the adsorbed H atoms will be delocalized and form a quasi two-dimensional gas.

Since OH_{ad} can also be experimentally identified, the mechanism of this reaction was believed to proceed along the steps (1)–(5), as shown in Fig. 6.25. However, the actual mechanism is more complicated [81]. H_2O desorbs at about 170K [82], and the net activation energy for the reaction was found to be much smaller below this temperature (0.13 eV) [83] than above room temperature ($\sim$0.7 eV) [84]. In addition, at low temperatures, an induction period was found [83]. This problem could be solved by combined application of STM and vibrational spectroscopy [81]. It turned out that reaction step (6), $H_2O_{ad} + O_{ad} \rightarrow 2OH_{ad}$, is very important as long as adsorbed H_2O is present, that is, below 170K in UHV.

The progress of the reaction is illustrated in the STM images of Fig. 6.26. If an O-covered Pt(1 1 1) surface is exposed to a constant H_2 pressure of 8×10^{-9} mbar at 131K, the O atoms appearing as dark dots are continuously replaced by bright patches with internal structure that are identified with ordered OH_{ad} islands. In the further progress of the reaction, the latter are transformed into fuzzy patterns of adsorbed H_2O, which then desorb. These

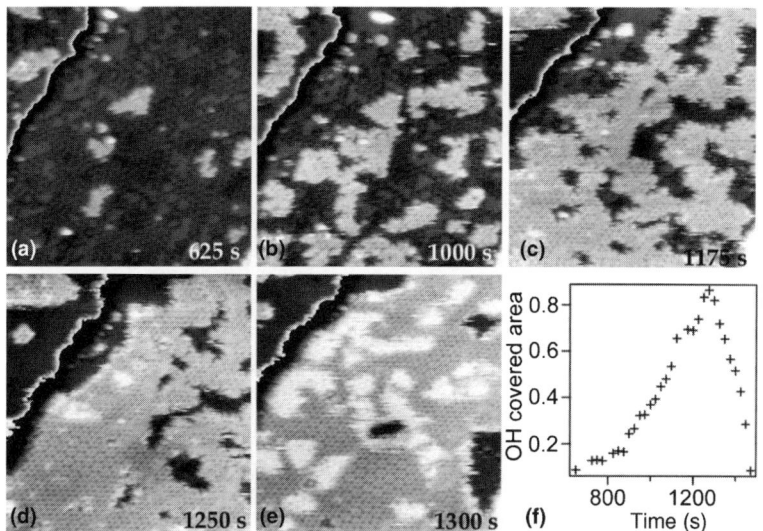

FIGURE 6.26. A series of STM images showing the progress of reaction between O atoms adsorbed on Pt(1 1 1) and H_2 (a–e). Fraction of the surface covered by OH_{ad} as a function of time (f) [81].

atomic scale data thus suggest operation of reaction steps (1)–(5). Reaction (6), however, readily takes place whenever O_{ad} and H_2O_{ad} meet and is indeed much faster than the formation of OH_{ad} through direct recombination of O_{ad} and H_{ad}. The actual situation is illustrated in the STM images of Fig. 6.27 recorded on a larger scale [85]. In the upper part, the bright spots are small patches of OH_{ad} that form a circular region with higher density. The latter expands continuously with time into the O-covered region leaving a blank area (where the H_2O formed had desorbed) behind. This process of wave propagation is schematically sketched in the lower part of Fig. 6.27: At the boundary between H_2O_{ad} and O_{ad}, formation of OH_{ad} takes place, which in an autocatalytic step is again transformed into H_2O_{ad}. Thus, a wave of OH_{ad} is continuously propagating into the O_{ad} region and leaves the reaction product H_2O behind. The front position varies linearly with time with a velocity of about 15 nm/min while the width of the front remained unchanged (~17 nm) during the experiment.

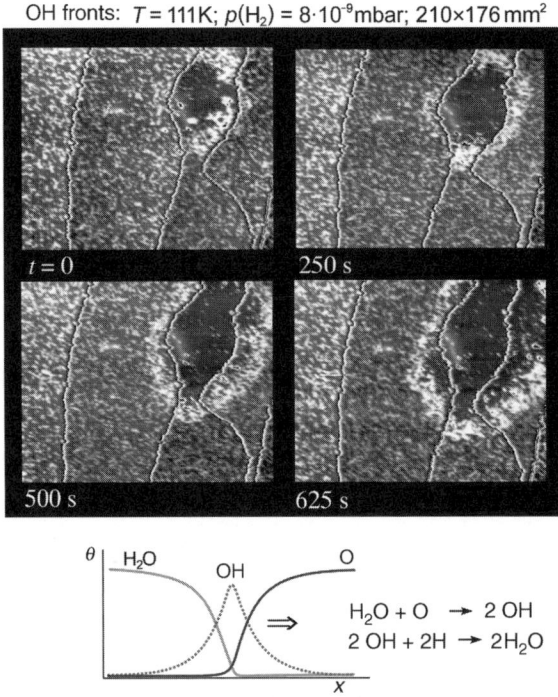

FIGURE 6.27. Large-scale STM images from the $H_2 + O_2$ reaction on Pt(1 1 1), illustrating the propagation of an OH_{ad} wave [85].

The simplified reaction scheme is plotted in Fig. 6.28, and theoretical modeling has to take into account the surface diffusion of adsorbed H_2O, apart from the kinetics, leading to the set of *partial* differential equations in the lower part of this figure. The result of numerical simulations [85,86] is reproduced in Fig. 6.29. The upper part shows the development of the 1D concentration profiles with time if a local nucleus of H_2O is placed into an O surrounding, while the lower part represents the 2D expansion of the OH ring. The length scale is no longer given by atomic dimensions, but by the so-called diffusion length determined by the combination of diffusion coefficient D and rate constant k_3, $l_D = \sqrt{D/k_3}$, which is of the order of µm. The figure shows that H_2O diffuses into the O-covered area where it produces OH according to the step k_3. The H_2O concentration of the

Reaction sequence: hydrogen oxidation on Pt(111)

$$O_{ad} + H_{ad} \xrightarrow{k_1} OH_{ad}$$
$$OH_{ad} + H_{ad} \xrightarrow{k_2} H_2O_{ad}$$
$$H_2O_{ad} + O_{ad} \xrightarrow{k_3} 2\, OH_{ad}$$

$$\frac{\partial}{\partial t}[H_2O] = -k_3[H_2O][O] + k_2[OH][H] + D\frac{\partial^2}{\partial t^2}[H_2O]$$

$$\frac{\partial}{\partial t}[OH] = 2k_3[H_2O][O] - k_2[OH][H]$$

$$\frac{\partial}{\partial t}[O] = -k_3[H_2O][O]$$

FIGURE 6.28. Reaction scheme for the $H_2 + O_2$ reaction on Pt(1 1 1) and resulting (simplified) partial differential equations (PDE's).

nucleus decreases and an OH peak develops at the O/H_2O interface that continuously propagates into the O_{ad} region. Thus, the calculations represent the experimental observations well, while quantitative comparison of the front propagation velocities

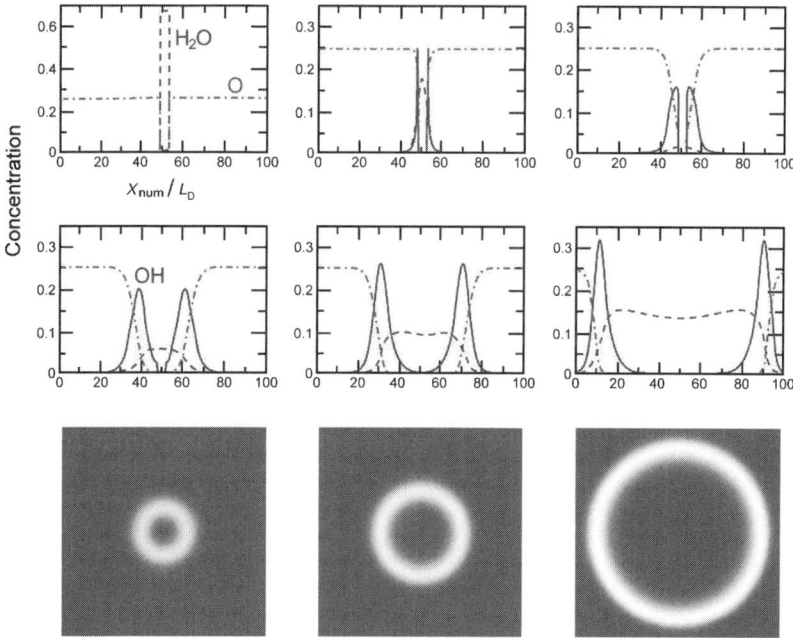

FIGURE 6.29. Numerical simulation of the formation and propagation of an OH_{ad} wave in the $H_2 + O_2$ reaction on Pt(1 1 1) [85,86]. (See color insert.)

differs and is attributed to the limitation of the applied mean field approximation of the reaction–diffusion (RD) model.

This system was discussed in detail because it demonstrates the limitations of the usually assumed uniform distribution of reacting species over the surface, where the temporal behavior may be described in terms of *ordinary* differential equations for the kinetics. The present reaction is characterized by nonlinear behavior and coupling of reaction and diffusion. Chapters 7 and 8 will be devoted to topics relating to such systems within the framework of nonlinear dynamics.

REFERENCES

1. W. Appl, *Ammonia*, Wiley-VCH, 1999.
2. J. R. Jennings (ed.), *Catalytic Ammonia Synthesis: Fundamentals and Practice*, Plenum Press, New York, 1991.
3. G. Ertl, in: *Encyclopedia of Catalysis* (eds. J. T. Horvath, E. Iglesia, M. T. Klein, J. A. Lercher, A. J. Russell, and G. I. Stiefel), Vol. **1**, Wiley, 2003, p. 389.
4. R. Schlögl, in: *Handbook of Heterogeneous Catalysis* (eds. G. Ertl, H. Knözinger, F. Schüth, and J. Weitkamp), Wiley, 2008, p. 2501.
5. A. Mittasch, *Geschichte der Ammoniaksynthese*, Verlag Chemie, Weinheim, 1957.
6. P. H. Emmett, in: *The Physical Basis for Heterogeneous Catalysis* (eds. R. I. Jaffee and E. Drauglis), Plenum, New York, 1975, p. 3.
7. G. Ertl, D. Prigge, R. Schlögl, and M. Weiss, *J. Catal.* **79** (1983) 359.
8. (a) G. Ertl, *Catal. Rev.* **21** (1980) 201; (b) G. Ertl, *J. Vac. Sci. Technol. A* **1** (1983) 1247.
9. P. H. Emmett and S. Brunauer, *J. Am. Chem. Soc.* **55** (1933) 1738; **56** (1934) 35.
10. N. D. Spencer, R. C. Schoonmaker, and G. A. Somorjai, *J. Catal.* **74** (1982) 129.
11. D. R. Strongin, J. Carrazza, S. Bare, and G. A. Somorjai, *J. Catal.* **103** (1987) 213.
12. J. A. Dumesic, H. Topsoe, and M. Boudart, *J. Catal.* **37** (1975) 513.
13. D. R. Strongin and G. A. Somorjai, *J. Catal.* **109** (1988) 51.
14. M. Grunze, M. Golze, W. Hirschwald, H. J. Freund, H. Pulm, U. Seip, M. C. Tsai, G. Ertl, and J. Küppers, *Phys. Rev. Lett.* **53** (1984) 850.
15. H. J. Freund, B. Bartos, R. P. Messmer, M. Grunze, H. Kuhlenbeck, and M. Neumann, *Surf. Sci.* **185** (1987) 187.

16. G. Ertl, S. B. Lee, and M. Weiss, *Surf. Sci.* **114** (1982) 515.
17. M. C. Tsai, U. Seip, I. Basssignana, J. Küppers, and G. Ertl, *Surf. Sci.* **155** (1985) 387.
18. R. Imbihl, R. J. Behm, G. Ertl, and W. Moritz, *Surf. Sci.* **123** (1982) 129.
19. M. Grunze, in: *The Chemical Physics of Solid Surfaces and Heterogeneous Catalysis* (eds. D. A. King and D. P. Woodruff), Vol. **4**, Elsevier, 1982, p. 143.
20. I. Alstrup, I. Chorkendorff, and S. Ullmann, *Z. Phys. Chem.* **198** (1997) 123.
21. C. Rettner and H. Stein, *Phys. Rev. Lett.* **59** (1987) 2768.
22. J. J. Mortensen, L. B. Hansen, B. Hammer, and J. K. Nørskov, *J. Catal.* **182** (1999) 479.
23. G. Ertl, S. B. Lee, and M. Weiss, *Surf. Sci.* **114** (1982) 527.
24. G. Ertl, M. Weiss, and S. B. Lee, *Chem. Phys. Lett.* **60** (1979) 391.
25. J. K. Nørskov, S. Holloway, and N. D. Lang, *Surf. Sci.* **137** (1984) 65.
26. Z. Paal, G. Ertl, and S. B. Lee, *Appl. Surf. Sci.* **8** (1981) 231.
27. D. R. Strongin and G. A. Somorjai, *Catal. Lett.* **1** (1988) 61.
28. F. Bozso, G. Ertl, M. Grunze, and M. Weiss, *Appl. Surf. Sci.* **1** (1977) 1.
29. M. Grunze, T. Bozso, G. Ertl, and M. Weiss, *Appl. Surf. Sci.* **1** (1978) 241.
30. M. Weiss, G. Ertl, and F. Nitschke, *Appl. Surf. Sci.* **2** (1979) 614.
31. M. Drechsler, H. Hoinkes, H. Kaarmann, H. Wilsch, G. Ertl, and M. Weiss, *Appl. Surf. Sci.* **3** (1979) 217.
32. M. I. Temkin and V. Pyzhev, *Acta Physicochim. URSS* **12** (1940) 489.
33. M. Boudart, *Ind. Chim. Belg.* **18** (1954) 489.
34. M. Boudart, *Catal. Lett.* **1** (1988) 21.
35. P. Stoltze and J. K. Nørskov, *Phys. Rev. Lett.* **55** (1985) 2502.
36. P. Stoltze and J. K. Nørskov, *J. Catal.* **110** (1988) 1.
37. M. Bowker, I. B. Parker, and K. C. Waugh, *Appl. Catal.* **14** (1985) 101.
38. I. B. Parker, M. Bowker, and K. C. Waugh, *J. Catal.* **114** (1988) 457.
39. J. A. Dumesic and A. A. Trevino, *J. Catal.* **116** (1989) 119.
40. H. Topsoe, M. Boudart, and J. K. Nørskov (eds.), *Topics in Catalysis: Frontiers in Catalytic Ammonia Synthesis and Beyond*, Vol. **1**, Balzers Science Publishers, 1994.
41. F. Haber, *Z. Elektrochem.* **16** (1910) 244.
42. K. I. Aika, H. Ori, and A. Ozaki, *J. Catal.* **27** (1972) 424.

43. (a) A. I. Forster, P. J. James, J. J. McCaroll, and S. R. Tennison,U.S. Patent 4,163,775 (Aug. 1979); (b) P. J. Shires, J. R. Cassata, B. G. Mandelik, and C. P. van Dijk, U.S. Patent 4,479,935 (Oct. 1984).
44. K. Honkala, A. Hellman, I. N. Remediakis, A. Logadottir, A. Carlsson, S. Dahl, C. H. Christensen, and J. K. Nørskov, *Science* **307** (2005) 555.
45. M. Muhler, F. Rosowski, O. Hinrichsen, A. Hornung, and G. Ertl, *Proceedings of the 11th International Congress on Catalysis*, Elsevier, 1996, p. 319.
46. O. Hinrichsen, F. Rosowski, M. Muhler, and G. Ertl, *Chem. Eng. Sci.* **51** (1996) 1683.
47. S. Dahl, A. Logadottir, R. C. Egeberg, J. H. Larsen, J. Chorkendorff, E. Törnqvist, and J. K. Nørskov, *Phys. Rev. Lett.* **83** (1999) 1814.
48. S. Bludau, M. Gierer, H. Over, and G. Ertl, *Chem. Phys. Lett.* **219** (1994) 452.
49. S. Schwegmann, A. P. Seitsonen, H. Dietrich, H. Bludau, H. Over, K. Jacobi, and G. Ertl, *Chem. Phys. Lett.* **264** (1997) 680.
50. J. J. Mortensen, B. Hammer, and J. K. Nørskov, *Phys. Rev. Lett.* **80** (1998) 4333.
51. A. Logadottir and J. K. Nørskov, *J. Catal.* **220** (2003) 273.
52. H. Dietrich, K. Jacobi, and G. Ertl, *J. Chem. Phys.* **105** (1996) 8944.
53. A. Hornung, M. Muhler, and G. Ertl, *Catal. Lett.* **53** (1998) 77.
54. T. Engel and G. Ertl, *Adv. Catalysis* **28** (1979) 1.
55. T. Engel and G. Ertl, in: *The Chemical Physics of Solid Surfaces and Heterogeneous Catalysis* (eds. D. A. King and J. P. Woodruff), Vol. 4, Elsevier, 1982, p. 73.
56. S. Schwegmann, H. Over, V. De Renzi, and G. Ertl, *Surf. Sci.* **375** (1997) 91.
57. H. Froitzheim, H. Hopster, H. Ibach, and S. Lehwald, *Appl. Phys.* **13** (1977) 197.
58. C. T. Campbell, G. Ertl, H. Kuipers, and J. Segner, *J. Chem. Phys.* **73** (1980) 5862.
59. K. H. Allers, H. Pfnür, P. Feulner, and D. Menzel, *J. Chem. Phys.* **100** (1994) 3985.
60. A. Alavi, P. Hu, T. Deutsch, P. L. Silvestrelli, and J. Hutter, *Phys. Rev. Lett.* **80** (1998) 3650.
61. J. L. Gland and E. B. Kollin, *J. Chem. Phys.* **78** (1983) 963.
62. J. Wintterlin, S. Völkening, T. V. W. Janssens, T. Zambelli, and G. Ertl, *Science* **278** (1997) 1931.
63. G. Ertl, M. Neumann, and K. M. Streit, *Surf. Sci.* **64** (1977) 393.
64. N. W. Cant, P. C. Hicks, and B. S. Lennon, *J. Catal.* **54** (1978) 372.

65. H. I. Lee and J. M. White, *J. Catal.* **63** (1980) 261.
66. V. I. Savchenko, G. K. Boreskov, A. V. Kalinka, and A. N. Salanov, *Kinet. Catal.* **24** (1984) 983.
67. C. H. F. Peden and D. W. Goodman, *J. Phys. Chem.* **90** (1986) 1360.
68. H. Over, Y. D. Kim, A. P. Seitsonen, S. Wendt, E. Lundgren, M. Schmidt, P. Varga, A. Morgante, and G. Ertl, *Science* **287** (2000) 1474.
69. C. Y. Fan, J. Wang, K. Jacobi, and G. Ertl, *J. Chem. Phys.* **114** (2001) 10058.
70. K. Reuter and M. Scheffler, *Phys. Rev. B* **68** (2003) 045407.
71. J. Wang, C. Y. Fan, K. Jacobi, and G. Ertl, *J. Phys. Chem. B* **106** (2002) 3422.
72. K. Reuter, D. Frenkel, and M. Scheffler, *Phys. Rev. Lett.* **93** (2004) 116105.
73. J. Assmann, V. Narkhede, N. A. Brauer, M. Muhler, A. P. Seitsonen, M. Knapp, D. Crihan, A. Farkas, G. Mellau, and H. Over, *J. Phys.* **20** (2008) 184017.
74. J. Assmann, D. Crihan, M. Knapp, E. Lundgren, E. Löffler, M. Muhler, V. Narkhede, H. Over, M. Schmid, A. P. Seitsonen, P. Varga, *Angew. Chem. Int. Ed.* **44** (2005) 917.
75. J. W. Döbereiner, *Schweigg. J.* **39** (1823) 1.
76. J. J. Berzelius, *Jber. Chem.* **15** (1837) 237.
77. G. Källen and G. Wahnström, *Phys. Rev. B* **65** (2001) 033406.
78. K. Nobuhara, H. Kasai, H. Nakaniski, and A. Okiji, *Surf. Sci.* **507** (2002) 82.
79. S. C. Badescu, P. Salo, T. Ala-Nissila, S. C. Ying, K. Jacobi, Y. Wang, K. Bedürftig, and G. Ertl, *Phys. Rev. Lett.* **88** (2002) 136101.
80. S. C. Badescu, K. Jacobi, Y. Wang, K. Bedürftig, G. Ertl, P. Salo, T. Ala-Nissiland, and S. C. Ying, *Phys. Rev. B* **68** (2003) 205401.
81. S. Völkening, K. Bedürftig, K. Jacobi, J. Wintterlin, and G. Ertl, *Phys. Rev. Lett.* **83** (1999) 3672.
82. G. B. Fisher and J. L. Gland, *Surf. Sci.* **94** (1980) 446.
83. K. M. Ogle and J. M. White, *Surf. Sci.* **139** (1984) 43.
84. A. B. Anton and D. C. Cadogan, *Surf. Sci.* **239** (1990) L548.
85. C. Sachs, M. Hildebrand, S. Völkening, J. Wintterlin, and G. Ertl, *Science* **293** (2001) 1635.
86. C. Sachs, M. Hildebrand, S. Völkening, J. Wintterlin, and G. Ertl, *J. Chem. Phys.* **116** (2002) 5759.

CHAPTER 7

OSCILLATORY KINETICS AND NONLINEAR DYNAMICS

7.1. INTRODUCTION

Erwin Schrödinger presented in 1943 a series of lectures that were afterward published as a book entitled *What is Life?* [1]. This book is considered as a milestone in the development of molecular biology, although the original intention of the author was somewhat different. He wanted to answer the question if the known laws of physics are able to explain why a biological system may develop spontaneously into a state of higher order, but he could not give a clear solution of this problem. The formation of structures that are ordered in space and/or time seems to be at variance with the second law of thermodynamics, whereafter all processes in a closed system (without attractive interactions between the constituents) tend to increase the entropy, that is, the grade of disorder. However, chemical reactions at steady-state flow conditions occur in open systems where a constant flow of free energy keeps the system far away from thermodynamic equilibrium. As a consequence, such systems may exhibit oscillatory or even chaotic behavior of the reaction

---
Reactions at Solid Surfaces. By Gerhard Ertl
Copyright © 2009 John Wiley & Sons, Inc.

rates, as well as the formation of spatiotemporal concentration patterns. The theoretical background for these phenomena was laid by Prigogine [2] (who denoted them as "dissipative structures") and by Haken [3] (in the framework of synergetics). This chapter will discuss some observations of temporal order without taking spatial structure formation into account, which will be the subject of Chapter 8.

Oscillatory kinetics with a surface reaction had been observed as early as in 1828 by Fechner [4] with an electrochemical system. As an example for these types of reactions, Fig. 7.1 shows the variation of the potential at a Pt electrode with time for the electrochemical oxidation of H_2 in the presence of copper ions [5]. While the potential at low-current density j is constant (a), at higher j kinetic oscillations occur because of periodic poisoning and activation transitions of the electrode by underpotential deposition and dissolution of a passivating Cu overlayer. With further increase of j, at first period doubling and then transition to an irregular situation (chaos) take place.

The rich variety of this type of temporal self-organization was mainly explored in detail with the famous Belousov–Zhabotinsky reaction [6,7], but with heterogeneously catalyzed reactions the oscillatory kinetics were first reported only around 1970 for the oxidation of CO on Pt catalysts [8,9]. Since then oscillatory kinetics have been found with more than a dozen catalytic reactions, and this field has also been extensively reviewed [10,17].

Chemical oscillators are described on the basis of nonlinear dynamics, in that the underlying kinetic equations under steady-state conditions are nonlinear. If one assumes that the spatial distribution of the reaction species is uniform, then these variables will only depend on time, and mathematical description in the mean field approximation for the concentration variables c_i is achieved by a set of coupled (nonlinear) ordinary differential equations (ODEs). This will be the approach applied in this chapter.

INTRODUCTION

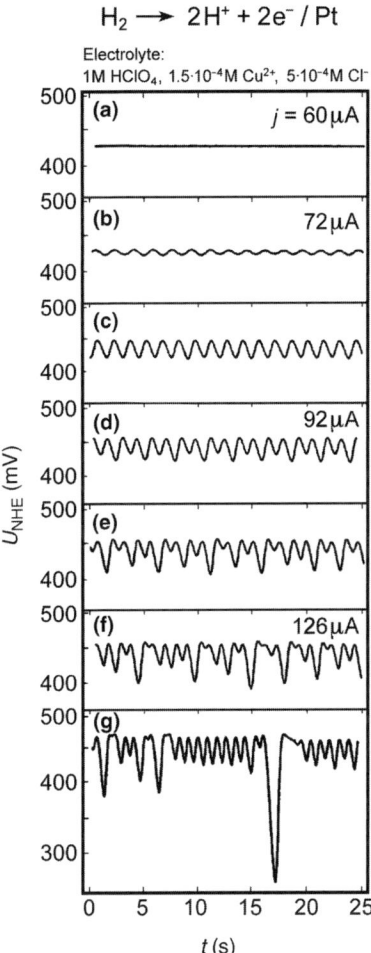

FIGURE 7.1. Time series of the potential of a Pt electrode during electrochemical oxidation of H_2 in the presence of copper ions at different current densities [5].

Generally, however, coupling between different parts of the surface is necessary to establish oscillations of the macroscopic kinetics. Hence, the concentration variables not only depend on time but also on spatial coordinates, and the dynamic behavior is now to be formulated in terms of partial differential equations (PDEs). As a consequence, spatiotemporal pattern formation takes place, which effects will be discussed in Chapter 8.

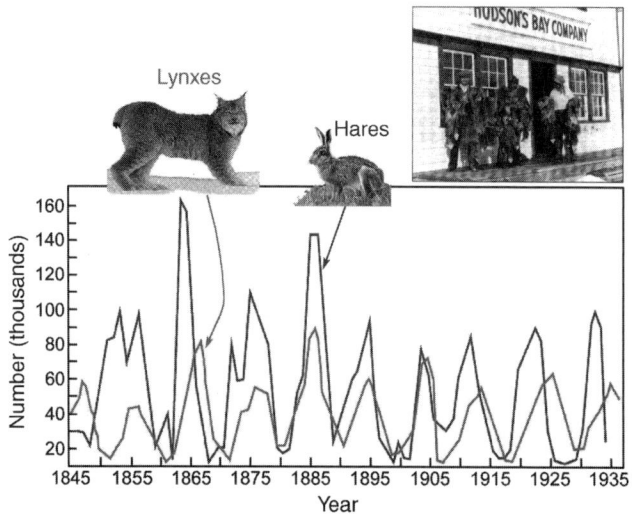

FIGURE 7.2. Variation with time of the number of furs (in thousands) from hares and lynxes delivered to the Hudson's Bay Company [18]. (See color insert.)

An example from population dynamics is presented in Fig. 7.2 that shows the variation with time of the number of furs from hares and lynxes delivered to the Hudson's Bay Company [18]. The oscillatory populations of both species are obviously coupled to each other with a certain phase shift. There is a plausible explanation: When the lynxes find enough food (i.e., hares) their population increases, while that of the hares decays as soon as the birth rate cannot compensate the growing loss anymore. When the supply of hares drops, the lynxes begin to starve and their population decreases, so the population of the hares recover again. An approximate mathematical description of this effect can be achieved in terms of two coupled, nonlinear ODEs (Lotka–Volterra model) [19] for the concentrations of hares, x, and lynxes, y, as presented in Fig. 7.3, together with the solution for properly chosen parameters α and β.

Systems of coupled differential equations of the aforementioned type may be subject to systematic analysis on the basis of

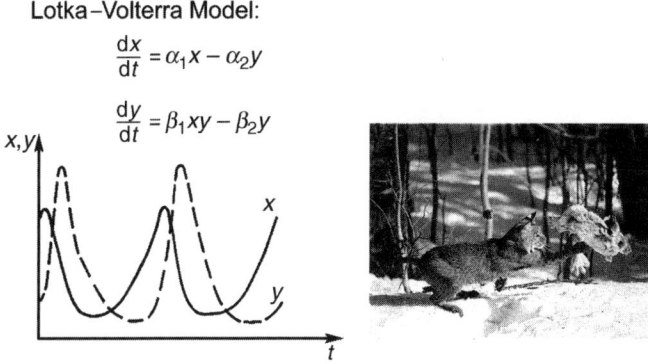

FIGURE 7.3. The Lotka–Volterra model with $x \hat{=}$ concentration of hares and $y \hat{=}$ concentration of lynxes.

bifurcation theory [20,21]. Bifurcation denotes a qualitative change of the dynamics upon variation of one of the control parameters, such as a transition from stationary to oscillatory behavior, as in Fig. 7.1 going from (a) to (b). The dynamical behavior of a system described by two variables can be either stationary or oscillatory, corresponding to fixed points or limit cycles, respectively, in a phase–space representations. Various mathematical models have been proposed to model oscillatory catalytic reactions. The decision, if a proposed reaction scheme provides oscillatory solutions, may be reached by integration of the corresponding differential equations or, even more elegantly, by applying a general strategy denoted as stoichiometric network analysis (SNA) [22].

In the following, the effects found with the oxidation of CO on a Pt(1 1 0) surface under isothermal conditions will be discussed in detail.

7.2. Oscillatory Kinetics in the Catalytic CO Oxidation on Pt(1 1 0)

The mechanism of the catalytic oxidation of CO on Pt surfaces was discussed in Section 6.3. This was also the first system exhibiting

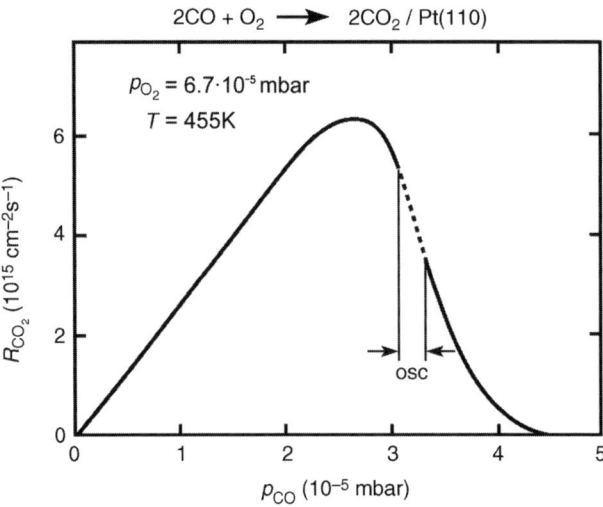

FIGURE 7.4. Steady-state rate of CO_2 formation in the catalytic oxidation of CO on a Pt(1 1 0) surface as a function of p_{CO} at fixed p_{O_2} and T.

kinetic oscillations to be investigated using the "surface science" approach, that is, with single-crystal surfaces under low pressure and hence isothermal conditions [23]. Figure 7.4 shows the steady-state rate of CO_2 formation on a Pt(1 1 0) surface as a function of CO pressure, while the other two control parameters, p_{O_2} and T, were kept constant. At low p_{CO}, the surface is largely covered by O_{ad} and the reaction rate is determined by the supply of CO, and therefore rises proportional to p_{CO} until the stationary CO coverage becomes so high that it begins to inhibit oxygen adsorption. Any further increase of p_{CO} causes a decrease of the reaction rate, which is now limited by oxygen adsorption. This decrease of rate becomes steeper at lower temperatures, and the system eventually becomes bistable, exhibiting branches of low and high reactivity and a clockwise hysteresis upon variation of p_{CO} [24], which can readily be modeled with a two-variable model [25]. While such a behavior is characteristic of the densely packed Pt(1 1 1) surface, novel effects, namely, rate oscillations, come into play with the more open crystal planes such as (1 1 0). Figure 7.5 shows how the

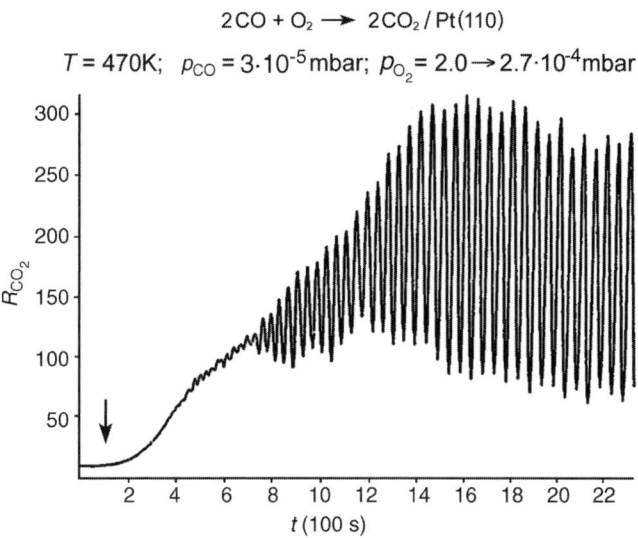

FIGURE 7.5. Onset of kinetic oscillations in the rate of CO oxidation at a Pt(1 1 0) surface [26].

rate varies if at the time marked by an arrow the O_2 pressure is increased stepwise from 2.0 to 2.7×10^{-4} mbar [26]. The rate increases slowly and then becomes oscillatory with constant amplitude. These oscillations occur within a narrow range of parameters as marked in Fig. 7.4, where the structure of the Pt (1 1 0) surface varies periodically between the reconstructed 1 × 2 and the nonreconstructed 1 × 1 phases. These two modifications of the surface structure were illustrated in Fig. 2.22. As outlined there, the clean Pt(1 1 0) surface is reconstructed into the 1 × 2 "missing row" structure that is transformed into the normal 1 × 1 structure if the CO coverage exceeds a critical coverage of about 0.2. This transformation is driven by the difference in CO adsorption energy on both phases. On the other hand, the sticking coefficient for dissociative oxygen adsorption is larger on the 1 × 1 than on the 1 × 2 phase. Since the kinetic oscillations are occurring under conditions for which oxygen adsorption is rate limiting, their origin may now be readily rationalized: Starting with the 1 × 2 surface, the $p_{O_2} : p_{CO}$ ratio will be such that enough CO is adsorbed to lift the reconstruction. On the 1 × 1 patches formed, more O will be adsorbed, which then reacts with adsorbed CO, so the coverage of the latter decreases below the critical value for keeping the 1 × 1 phase stable. The surface structure transforms back to 1 × 2 where the oxygen sticking coefficient is smaller, so the CO coverage could build up again and the cycle may be repeated. Mathematical modeling of the kinetics may now be achieved on the basis of three variables for the coverages of CO and O, u and v, respectively, and the fraction w of the surface being present as 1 × 1 phase, taking into account the difference of the O_2 sticking coefficient between the two phases and the dependence of the sticking coefficients for both CO and O on the respective coverages [31]. The resulting system of ODEs is reproduced in Fig. 7.6, and the solutions for the rate as well as for the three coverages as function of time for a specific set of control parameters are shown in Fig. 7.7. Note that the O and CO coverages are strictly anticorrelated, while the fraction of the surface existing as 1 × 1 phase oscillates with a certain phase shift and plays the role of a slow variable whose time constant is governed by the kinetics of the structural phase transformation.

$$2CO + O_2 \rightarrow 2CO_2 / Pt(110)$$

Mechanism:

$CO + * \rightleftarrows CO_{ad}$

$O_2 + 2* \rightarrow 2O_{ad}$

$O_{ad} + CO_{ad} \rightarrow CO_2 + 2*$

$Pt(1\times2) \xleftarrow{CO_{ad}} (1\times1)$

Kinetic modeling:

$\theta_{CO} = u; \theta_O = v; \theta_{1\times1} = w(\theta_{1\times2} = 1-w)$

$$\frac{du}{dt} = s(CO)p_{CO} - k_2 u - k_3 uv \quad (7.1)$$

$$\frac{dv}{dt} = s(O_2)p_{O_2} - k_3 uv \quad (7.2)$$

$$\frac{dw}{dt} = k_5[f(u) - w] \quad (7.3)$$

$$s(CO) = k_1(1 - u^3)$$

$$s(O_2) = k_4[s_1 w + s_2(1-w)](1-u-v)^2$$

FIGURE 7.6. Kinetic modeling of the rate oscillations in CO oxidation on Pt (1 1 0) [25].

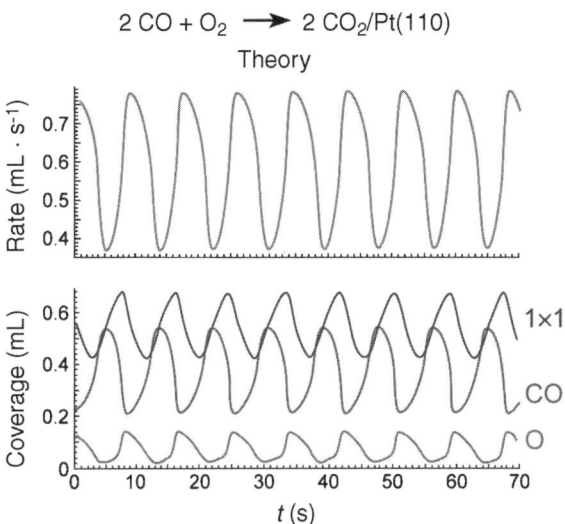

FIGURE 7.7. Solution of the set of ordinary differential equations presented in Fig. 7.6 and describing the kinetic oscillations in CO oxidation on Pt(1 1 0) [25]. (See color insert.)

The reaction rate strictly parallels the O coverages, as expected. The latter quantity is reflected by the change of the work function $\Delta\varphi$, which therefore may serve as a convenient experimental monitor for the rate oscillations, rather than by the direct recording of the CO_2 partial pressure by a quadrupole mass spectrometer.

Figure 7.8 shows a time series that was measured in this way, where p_{O_2} and T were kept constant, while p_{CO} was varied stepwise at the points marked by arrows. The oscillations are at first strictly periodic, but at a certain value of p_{CO} (1.64×10^{-4} mbar) a qualitative change (i.e., a bifurcation) takes place. Alternating large and small amplitudes are now formed and the period is doubled. Upon further decrease of p_{CO} the amplitude ratio varies, until at $p_{CO} = 1.60 \times 10^{-4}$ mbar another period doubling takes place. Very little variation of p_{CO} to 1.59×10^{-4} causes another qualitative change. Now the time series became irregular,

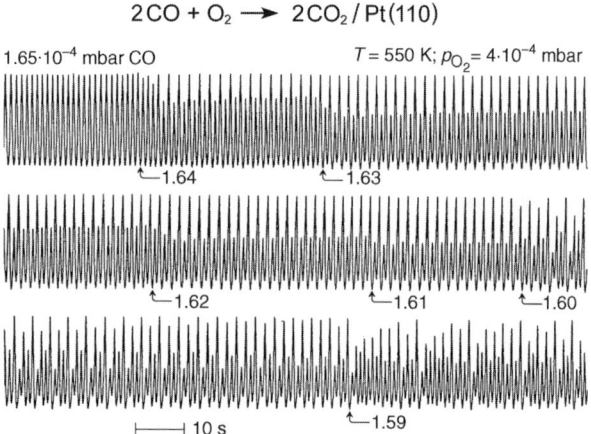

FIGURE 7.8. Kinetic oscillations in the CO oxidation on Pt(1 1 0) exhibiting the period doubling (Feigenbaum) transition to chaos [27].

a state denoted as deterministic chaos [27]. This transition to chaos via a sequence of period doublings is the well-known Feigenbaum scenario. The time series may be transformed into the corresponding phase portraits by the time delay method sketched in Fig. 7.9 [28]. The resulting phase portraits for these data are reproduced in Fig. 7.10 and exhibit the transition from a simple limit cycle to a "strange attractor." More detailed analysis revealed that for the chaotic state, one of the so-called Lyapunov exponents became positive, so the initially nearby trajectories in phase space became divergent as characteristic of the chaotic state [27]. The analysis of another set of data further inside the

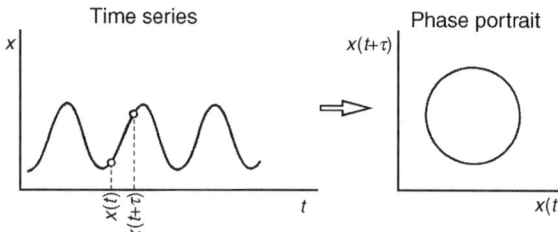

FIGURE 7.9. Construction of the phase portrait of a dynamic system from its time series [28].

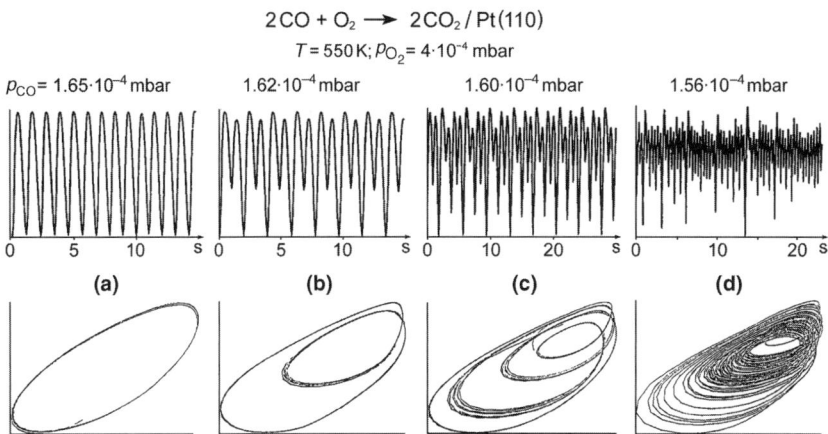

FIGURE 7.10. Phase portraits for the transition to chaos of the data presented in Fig. 7.8 [27].

chaotic window revealed that even the two Lyapunov exponents became positive, a situation denoted as "hyperchaos" [29].

Mathematical modeling in terms of the three ODEs already presented reproduces the stationary kinetics as well as simple periodic oscillations, but fails to describe the period doubling sequence leading to chaos [25]. This is due to the fact that this approximation assumes laterally uniform concentrations of the surface species where the whole surface is strongly synchronized so that these concentrations depend only on time. This is in fact not fulfilled, and spatial coupling leads to the formation of propagating concentration patterns, which will be the subject of Chapter 8. In addition, full theoretical description has taken into account the formation of subsurface oxygen species as a further variable [30].

7.3. FORCED OSCILLATIONS IN CO OXIDATION ON Pt(1 1 0)

A system exhibiting autonomous oscillations with frequency ν_o may in addition become subject to external periodic forcing of one of its control parameters where new dynamic features may arise.

The response to external forcing with frequency ν_p and amplitude A may be classified as follows [31–33]: If the resulting period T_r of the system exhibit a fixed phase relation to that of the modulation T_{ex}, the system is entrained. The ratio T_r/T_{ex} may be expressed as that between two small numbers, that is, $T_r/T_{ex} = k/l$. For $k/l = 1$, the entrainment is called harmonic, for $k/l > 1$ superharmonic, and for $k/l < 1$ subharmonic. If the phase difference between response and modulation varies continuously, the oscillations are called quasi-periodic.

The behavior of such a system may be characterized by a dynamic phase diagram, in which the regions of entrainment and quasi-periodicity are identified as a function of T_{ex}/T_o and A. The structure of such a diagram does not depend on the specific properties of the reacting system, but rather on its type of bifurcation around which the perturbation is applied [32,33].

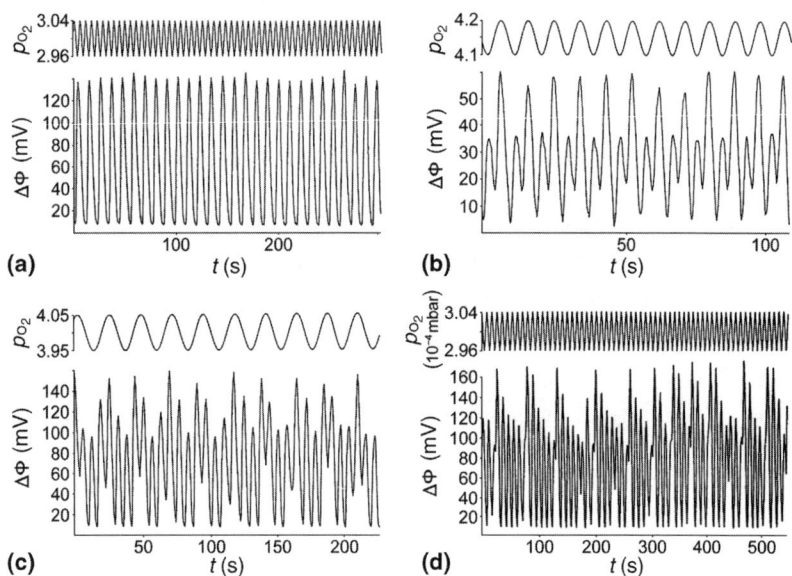

FIGURE 7.11. Time series for periodically forced oscillations in the CO oxidation on a Pt(1 1 0) surface exhibiting sustained oscillations [34]. (a) 1:2 Subharmonic entrainment. (b) 2:1 Superharmonic entrainment. (c) 7:2 Superharmonic entrainment. (d) Quasi-periodic response.

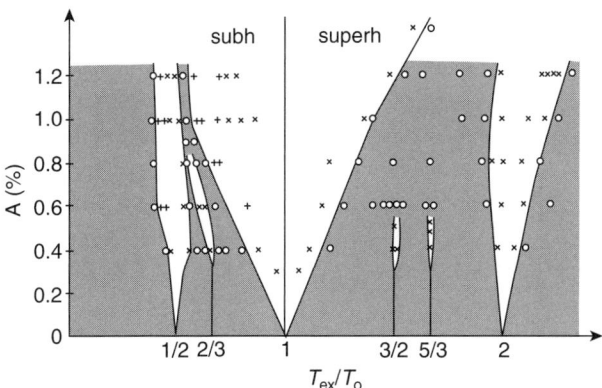

FIGURE 7.12. Experimental dynamic phase diagram for periodically forced oscillations in the CO oxidation on a Pt(1 1 0) surface [34]. Existence range for entrained and quasi-periodic (shaded areas) oscillations as a function of the ratio of the periods of modulation T_{ex} and autonomous oscillations T_o and amplitude A of the modulated O_2 pressure.

Figure 7.11 shows several time series for forced oscillations with the system CO oxidation on Pt(1 1 0) [34], where the partial pressure of O_2 was periodically modulated with frequency ν_p and amplitude A (as percentage of the basic value), and the response of the system recorded via the change of the work function $\Delta\varphi$ (which is proportional to the reaction rate). The resulting dynamic phase

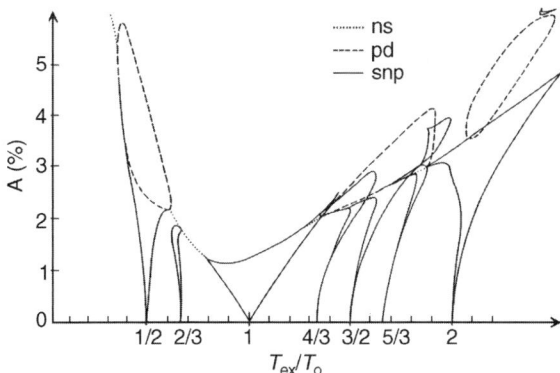

FIGURE 7.13. Theoretical dynamic phase diagram resulting from solution of the quoted ODEs [35]. Types of bifurcations: Ns = Neimark–Sacker, pd = period doubling, and snp = saddle node.

diagram is reproduced in Fig. 7.12 and is in good agreement with the theoretical results obtained from solution of the above-outlined ODEs with the respective forcing parameters as shown in Fig. 7.13 [35].

REFERENCES

1. E. Schrödinger, *What is Life?* Cambridge University Press, 1944.
2. G. Nicolis and I. Prigogine, *Self-Organization in Non-Equilibrium Systems*, Wiley, 1977.
3. (a) H. Haken, *Synergetics: An Introduction*, Springer, 1977; (b) A. S. Mikhailov, *Foundations of Synergetics*, Vols. I and II, Springer, 1991.
4. G. Th. Fechner, *Schweigg. J.* **53** (1828) 61.
5. K. Krischer, M. Lübke, W. Wolf, M. Eiswirth, and G. Ertl, *Ber. Bunsenges.* **95** (1991) 820.
6. R. J. Field and M. Burger (eds.), *Oscillations and Traveling Waves in Chemical Systems*, Wiley, 1985.
7. A. M. Zhabotinsky, *Ber. Bunsenges.* **84** (1980) 334.
8. P. Hugo, *Ber. Bunsenges.* **74** (1970) 121.
9. W. Beusch, D. Fieguth, and E. Wicke, *Chem. Ing. Tech.* **44** (1972) 445.
10. M. Sheintuch and R. A. Schmitz, *Catal. Rev.* **15** (1977) 107.
11. M. G. Slinko and M. M. Slinko, *Catal. Rev.* **17** (1978) 119.
12. G. Ertl, *Adv. Catal.* **37** (1990) 213.
13. F. Schüth, B. E. Henry, and L. D. Schmidt, *Adv. Catal.* **39** (1993) 51.
14. M. Slinko and N. Jaeger, *Oscillating Heterogeneous Catalytic Systems*, Elsevier, 1994.
15. R. Imbihl and G. Ertl, *Chem. Rev.* **95** (1995) 697.
16. M. Slinko and N. Jaeger(eds.), *Catal. Today* **105** (Special Issue) (2005) 181.
17. G. Ertl, in: *Handbook of Heterogeneous Catalysis* (eds. G. Ertl, H. Knözinger, F. Schüth, and J. Weitkamp), Wiley, 2008, p. 1492.
18. D. A. McLulich, *Variations in the Number of Hare*, University of Toronto Press, Toronto, 1937, after Ref. [3a].
19. A. J. Lotka, *Analytical Theory of Biological Populations*, Springer, 1998.
20. J. Guckenheimer and P. Holmes, *Nonlinear Dynamics, Dynamical Systems, and Bifurcation of Vector Fields*, Springer, 1986.

References

21. J. M. T. Thompson and H. B. Stewart, *Nonlinear Dynamics and Chaos*, Wiley, 1987.
22. M. Eiswirth, A. Freund, and J. Ross, *J. Phys. Chem.* **95** (1991) 1294.
23. G. Ertl, P. R. Norton, and J. Rüstig, *Phys. Rev. Lett.* **49** (1982) 177.
24. M. Ehsasi, S. Rezaie-Serej, J. H. Block, and K. Christmann, *J. Chem. Phys.* **92** (1990) 4949.
25. K. Krischer, M. Eiswirth, and G. Ertl, *J. Chem. Phys.* **96** (1992) 9161.
26. M. Eiswirth and G. Ertl, *Surf. Sci.* **177** (1986) 90.
27. M. Eiswirth, K. Krischer, and G. Ertl, *Surf. Sci.* **202** (1988) 565.
28. F. Takens, in: *Dynamical Systems and Turbulence* (eds. D. A. Rond and L. S. Young), Springer, 1981, p. 336.
29. M. Eiswirth, T. M. Kruel, G. Ertl, and F. W. Schneider, *Chem. Phys. Lett.* **193** (1992) 305.
30. A. von Oertzen, H. H. Rotermund, A. S. Mikhailov, and G. Ertl, *J. Phys. Chem. B* **104** (2000) 3155.
31. W. Vance, G. E. Tsahouras, and J. Ross, *Progr. Theor. Phys.* **99** (1989) 331.
32. G. E. Tsahouras and J. Ross, *J. Chem. Phys.* **87** (1987) 6538; **89** (1989) 5715.
33. G. E. Tsahouras and J. Ross, *J. Phys. Chem.* **93** (1989) 2853.
34. M. Eiswirth and G. Ertl, *Phys. Rev. Lett.* **60** (1988) 1526.
35. K. Krischer, M. Eiswirth, and G. Ertl, *J. Chem. Phys.* **97** (1992) 307.

CHAPTER 8

SPATIOTEMPORAL SELF-ORGANIZATION IN SURFACE REACTIONS

8.1. INTRODUCTION

The term "self-organization" to characterize the formation of order in chemical systems is used in two different ways:

1. *Closed systems* exhibit a tendency to reach equilibrium by lowering their free energy. The result is spatial ordering on *microscopic* (i.e., atomic) scale, where kinetic constraints may prevent the attainment of full equilibrium. The formation of ordered phases as a consequence of interactions between the surface species has been demonstrated with numerous examples in the preceding chapters. These effects are better denoted as *self-assembly*.

2. *Open systems far from equilibrium* will be the subject of this chapter. This situation is, for example, given for catalytic reactions under steady-state flow conditions. Apart from oscillatory or chaotic kinetics as described in Chapter 7, the interplay between reaction and transport processes may lead to the formation of concentration patterns on *mesoscopic*

Reactions at Solid Surfaces. By Gerhard Ertl
Copyright © 2009 John Wiley & Sons, Inc.

scale (typically >1 µm). These phenomena are considered as true *self-organization* in the framework of nonlinear dynamics.

Generally, an extended system such as a single-crystal surface or, even more, a supported catalyst that exhibits temporal variations of the macroscopic reaction rate must be subject to lateral coupling mechanisms between different parts. Otherwise, superposition of the uncorrelated contributions from different areas would average each other, resulting in a time-independent reaction rate. As a consequence, the coverages of the adsorbates will in general depend not only on time but also on the spatial coordinates of the surface, and proper mathematical modeling has to be performed in terms of partial differential equations (PDEs). Thus, the following coupling mechanisms may be operating:

(i) *Surface diffusion.* Local differences of the adsorbate coverages Θ_i will lead to surface diffusion, and the resulting reaction–diffusion (RD) equations will be of the type [1]

$$\frac{\partial \Theta_i}{\partial t} = f_i(\Theta_j, \lambda_k) + D_i \Delta C_i \qquad (8.1)$$

f_i represent nonlinear functions describing the kinetics, λ_k are the external control parameters, D_i is the diffusion coefficient, and Δ is the Laplace operator. Without reaction, this lateral communication through diffusion will lead to uniform distributions, but in combination with certain types of nonlinear reaction terms, spatial concentration patterns may be formed. These types of RD patterns in different fields have been extensively studied in the past [2]. Diffusion lengths in catalytic surface reactions are typically of the order >1 µm and are hence observed only with extended single-crystal surfaces. However, other effects on the nm scale will come into play with supported catalyst particles or in experiments with field emission tips, as will be discussed later.

The transition from the atomic to the mesoscopic scale by coupling of reaction and diffusion had been demonstrated by the STM investigation of the catalytic oxidation of H_2 on Pt(1 1 1), as discussed in Section 6.4 where modeling with proper PDEs was also presented.

(ii) *Heat conductance.* Local differences of the surface reactivity will cause temperature differences due to the finite reaction enthalpies. Coupling between different regions will then occur through heat flux counterbalancing the temperature gradients. This mechanism will dominate with supported catalysts where substantial temperature differences may develop and coupling also occurs via the support material where surface diffusion of the adsorbates is inhibited. The characteristic transport length is in this case of the order of 1 mm, much larger than the separation between the different catalyst particles or even the sizes of their individual crystal planes. As a consequence, such a system may still be regarded to be uniform on this length scale, but the phenomena will be strongly affected by heat phenomena rather than by details of the surface chemistry, and external constraints such as a constant average temperature may be applied.

(iii) *Gas-phase coupling.* Varying reactivity will also affect the partial pressures of the reactants, even in a flow system. These effects will be of particular relevance in experiments with single crystals under low-pressure ($<10^{-4}$ mbar) conditions where the temperature changes will become negligible. In these cases, the mean free paths of gaseous molecules exceed even the dimensions of the vacuum vessel. Here the molecules propagate with a speed of the order 1000 m/s, so the concentration differences are transmitted practically instantaneously. This offers a rather efficient "global" coupling mechanism that synchronizes different parts of a single-crystal surface or even separated samples [3]. The experiments with periodic forcing of an oscillating surface reaction described in Section 7.3 demonstrated that indeed pressure variations far below 1% may suffice to cause such synchronization effects [4].

(iv) *Electric field.* In electrochemical systems the ions are affected by the electric field in the solution and spatial coupling occurs through migration rather than by diffusion. In this case, the interfacial or double-layer potential represents the central variable for the buildup of potential patterns at the surface. However, the control of the experiment by an electronic device may cause the dynamics of each location to depend on the average double-layer potential that leads to global coupling [5].

8.2. Turing Patterns and Electrochemical Systems

The first theoretical description of the formation of concentration patterns in a chemically reacting system was presented by Turing [6] as a model for morphogenesis. In this model, *stationary* (i.e., time independent) patterns may develop in an open system containing two reacting species, provided that their diffusion coefficients are sufficiently different. Since in solutions this is usually not the case, experimental verification of these Turing patterns was achieved rather late and with one specific type of chemical oscillator [7,8]. In these experiments, the necessary difference of the mobilities was achieved by using a macromolecular substance that partly immobilizes the critical species by reversible complexation. On the other hand, the Turing mechanism was recently made responsible for structure formation in a biological system [9].

In a theoretical work, it was suggested that a certain class of electrochemical systems might exhibit Turing-type structures in which the restriction given by sufficiently strong differences of the diffusion coefficient is overcome [10]. In these systems, the electrode potential (the "inhibitor") plays the role of the rapidly diffusing species for which migration in the electric field is the relevant transport process, while the chemical variable (the "activator") is subject to normal diffusion. The ratio of the characteristic rates of the transport process then becomes very large, so the conditions for the formation of Turing patterns should be fulfilled in systems with S-shaped current–potential (I–ϕ_{DL}) characteristics. This may be realized with a system in which a first-order phase transition of an organic adsorbate is coupled with faradaic reaction of some electroactive species.

Figure 8.1 sketches the basic mechanism underlying this type of stationary pattern formation [5]: The concentration c of the activator (red line) is locally slightly perturbed from its steady-state value c_{ss} (a). This initiates the self-enhanced production of the activator that now slowly diffuses into neighboring regions (b). As a consequence, the inhibitor, that is, the double-layer potential (ϕ_{DL}), grows (blue dashed line) that propagates more rapidly out of the activator spot (c). Since the inhibitor consumes the activator, the concentration of the latter will always be suppressed outside the spot (d). In this way, a stationary concentration profile may develop surrounded by a halo of increased inhibitor ϕ_{DL}.

Experimental verification of this principle could be achieved by the technique of surface plasmon microscopy by which the lateral distribution of the electrode potential at a thin film can be monitored [11]. The experiments were performed with the reduction of periodate (IO_4^-) in the presence of camphor on a thin, preferentially (1 1 1)-oriented Au film [12]. Adsorbed camphor exhibits two first-order phase transitions upon variation of the electrode potential leading to the required S-shaped current–voltage characteristics. (The addition of perchlorate to the

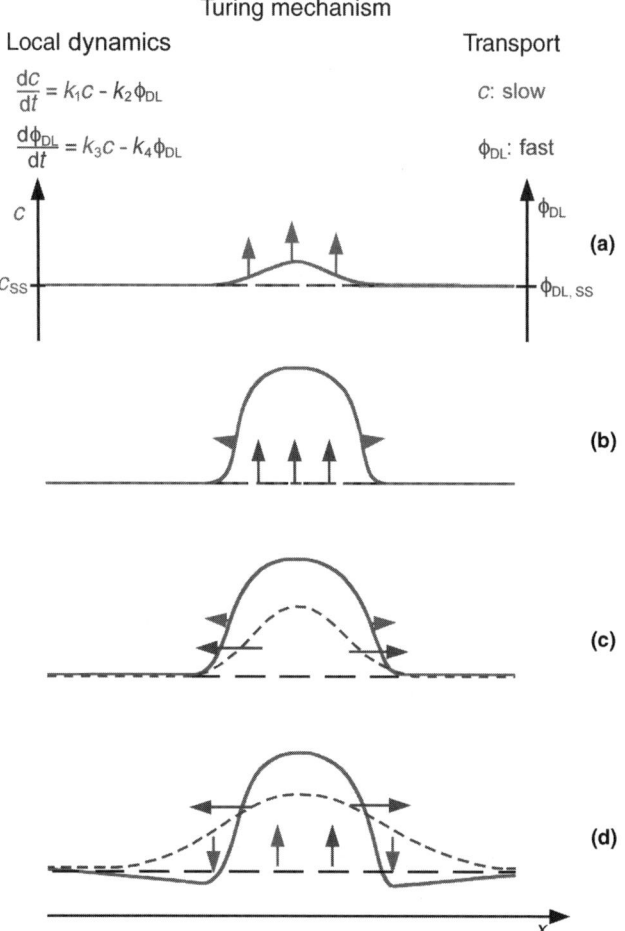

FIGURE 8.1. Principle of the mechanism underlying the formation of electrochemical Turing patterns [5]. (See color insert.)

solution served to modify the electrolyte resistivity.) Some of the resulting patterns are reproduced in Fig. 8.2. The length scale in these cases is of the order of 10 mm and may be varied by changing the reaction conditions. These patterns do not change with time if the parameters are kept constant and are hence of the Turing type. The presented mechanism can be verified with many organic adsorbates, and electrochemical membrane

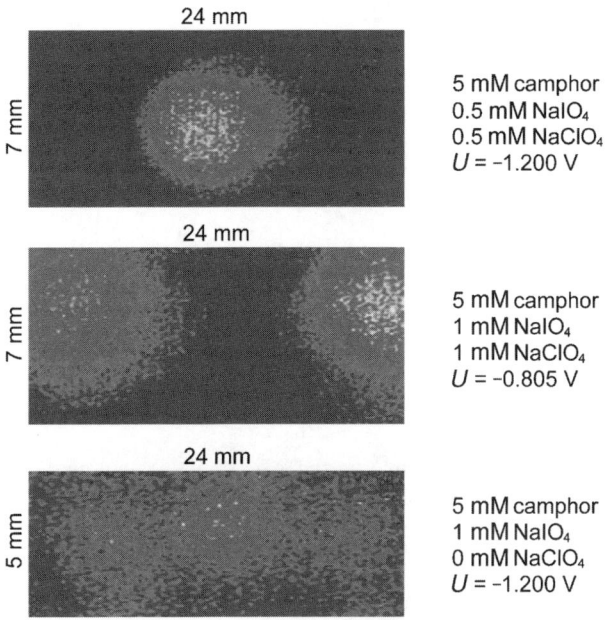

FIGURE 8.2. Some examples for the formation of electrochemical Turing patterns on a thin Au(1 1 1) film [12]. (See color insert.)

systems also exhibit first-order phase transitions [13] that may be used as models for biological membranes. It will be interesting to investigate if the same mechanism is responsible for the formation of structures in biological systems where potential gradients exist.

Apart from the stationary potential patterns just discussed, propagating potential waves under the influence of global (nonlocal) coupling via migration of ions in the electric field are much more readily realized in electrochemical systems [5]. These effects may be most conveniently studied with quasi-one-dimensional systems, that is, ring electrodes where the potential can be recorded at various locations. One example is concerned with the potentiostatic electrochemical oxidation of formic acid on a platinum ring electrode under bistable conditions, as

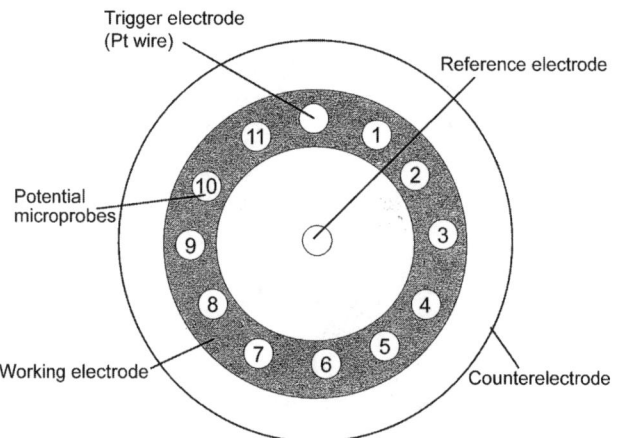

FIGURE 8.3. Schematic sketch of the electrode arrangement for studying remote triggering of potential waves in an electrochemical system [44].

sketched in Fig. 8.3 [44]. Here an appropriate perturbation at one location can cause the initiation of a wave on the opposite side of the ring (remote triggering). Some of the results are reproduced in Fig. 8.4 as space–time plots of the potential distribution along this ring electrode during passive–active transitions. Figure 8.4a represents an example for local triggering: A pulse of $+3\,\text{V}$ and $0.15\,\text{s}$ duration was applied to the electrode at

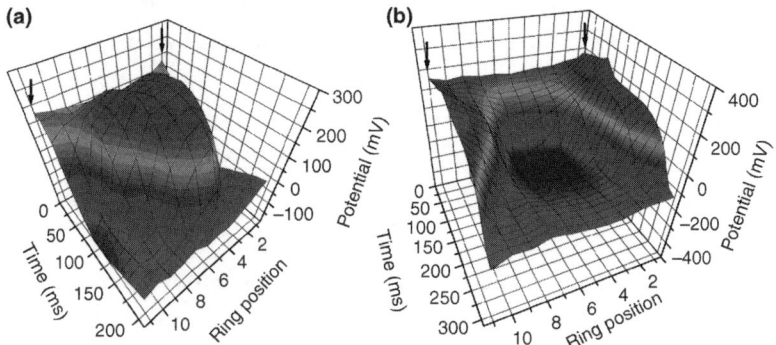

FIGURE 8.4. Space–time plots of the potential distributions on a Pt ring electrode during passive–active transitions. The ring position refers to Fig. 8.3 [44]. (a) Local triggering. (b) Remote triggering. (See color insert.)

position 12, triggering two propagating fronts at the neighboring positions 1 and 11. In Fig. 8.4b, on the other hand, perturbation with a pulse of $-5\,\text{V}$ and $0.1\,\text{s}$ duration causes the onset of activation at position 6 just opposite to the perturbation. These effects could be perfectly reproduced theoretically by taking the nonlocal character of migration coupling into account. The apparently instantaneous coupling arises from the slow timescale of the reaction (about $0.1\,\text{s}$ for the removal of the passivating OH layer), while the electric field spreads with the velocity of light. Such effects are also expected to play a role in biological systems where electric field effects come into play.

The use of a ribbon-shaped Pt electrode in electrooxidation of formic acid gives rise to edge effects of the interfacial potential, as is predicted from the potential theory in the form of the corresponding reaction–migration equation [45]. The edges tend to be more passive than the bulk of the electrode, which also causes a passivation transition to originate from the edges of the ribbon. In turn, an activation transition starts preferentially from its center. These effects are of relevance for various applications.

8.3. ISOTHERMAL WAVE PATTERNS

The $H_2 + O_2$ reaction on Pt(1 1 1), discussed in Section 6.4, had already demonstrated that a nonlinear reaction system may cause the formation of propagating concentration waves with typical length scales of $>1\,\mu\text{m}$, given by the diffusion length of the adsorbates. Imaging of these features may be achieved by photoemission electron microscopy (PEEM) [14] or by optical techniques [15].

The principle of PEEM is based on the different dipole moments of adsorbate complexes giving rise to modifications of the local work function. The yield of photoelectrons emitted from a surface irradiated by ultraviolet light is thus determined by the type and concentration of adsorbed species, and the lateral intensity distribution is imaged through a system of electrostatic lenses onto a channel plate and a fluorescent screen. From there the PEEM images are recorded by means of a charge coupled device (CCD) camera and stored on videotape. Typical resolutions are 0.9 μm and 20 ms.

Since the PEEM technique is based on the emission of electrons, its application is restricted to pressures below about 10^{-4} mbar. For higher pressures, two optical methods were adopted: ellipsometry for surface imaging (EMSI) and reflection anisotropy microscopy (RAM) [15].

The following discussion will concentrate on phenomena observed with our "drosophila," the catalytic CO oxidation on a Pt(1 1 0) surface. The contrast with the PEEM technique is based on local differences of the work function. Since adsorbed O causes a stronger increase of this property than adsorbed CO, areas on the surface predominantly covered by O_{ad} will appear dark in the images, while those primarily covered by CO_{ad} appear brighter.

A rich variety of spatiotemporal patterns on mesoscopic scale can thus be observed [16,17]. As an example, Fig. 8.5 presents a

$$2\,CO + O_2 \rightarrow 2\,CO_2/Pt(110)$$

$p_{O_2}= 3.2\cdot 10^{-4}$ mbar, $p_{CO} = 3\cdot 10^{-5}$ mbar, $T = 427$K, $\Delta t = 4.1$ s

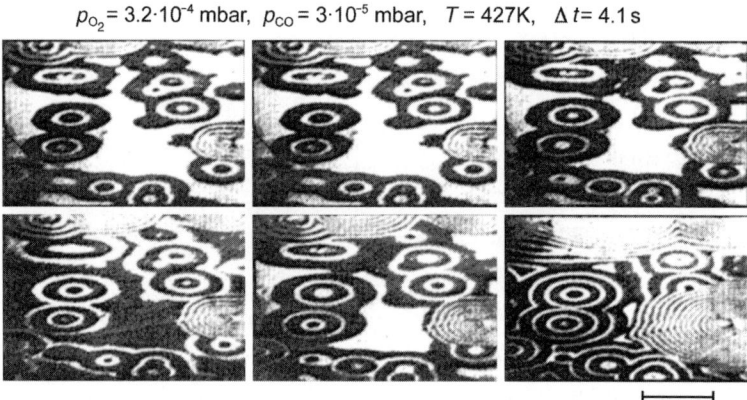

$\overline{100\ \mu m}$

FIGURE 8.5. A series of PEEM images from "target" patterns recorded during the steady-state reaction of CO oxidation on Pt(1 1 0) at intervals of 4.1 s [18].

series of "target" patterns, recorded with the indicated parameters at intervals of 4.1 s [18]. Concentric elliptical waves are propagating into a background periodically varying between dark (O) and bright (CO). The elliptical waves are elongated along the [1$\bar{1}$0] orientation. This demonstrates their origin as reaction–diffusion patterns. The periodically varying background reflects the oscillatory kinetics under these conditions and is caused by global coupling via the gas phase that mainly takes place by small variations of the CO pressure, since oxygen is present in large excess under these oscillatory conditions.

Theoretical modeling has now to include surface diffusion of the adsorbates, where this effect may be restricted to adsorbed CO as the fastest surface species. In the system of ODEs presented in Fig. 7.6, Eq. 7.1 has to now be supplemented by the additional term $D\Delta u$, where D is the diffusion coefficient of adsorbed CO with concentration u, and the resulting partial differential equations may then be solved numerically.

Rapid kinetic oscillations at higher temperatures, as reproduced in Fig. 7.8, are again synchronized by global coupling and are associated with standing waves, as reproduced in Fig. 8.6 [18]. These stripes are reminiscent of the patterns exhibited by zebras. Theoretical description of these effects was performed with the quoted set of PDEs by taking gas-phase coupling into account [19], but complete agreement with experiment could only be achieved later by inclusion of an additional equation for the concentration of subsurface oxygen [20].

The anisotropy of adsorbate diffusion with this system gives rise to a novel phenomenon, namely, the formation of solitary waves, as shown in Fig. 8.7 [21]. As indicated in the figure, these pulses propagate without changing their profile with a constant velocity of about 30 μm/s along the [0 0 1] direction at the given conditions. When two pulses traveling in opposite directions collide with each other, they are usually annihilated, as expected for RD systems, but sometimes one (or even two) of

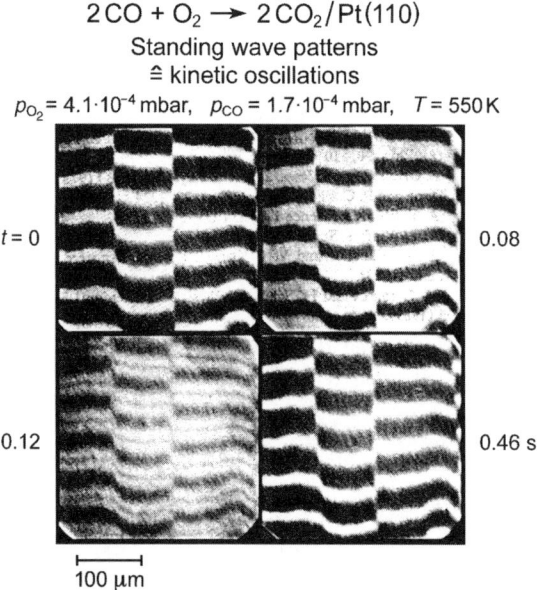

FIGURE 8.6. Standing wave patterns during sustained kinetic oscillations in the CO oxidation on Pt(1 1 0) [18].

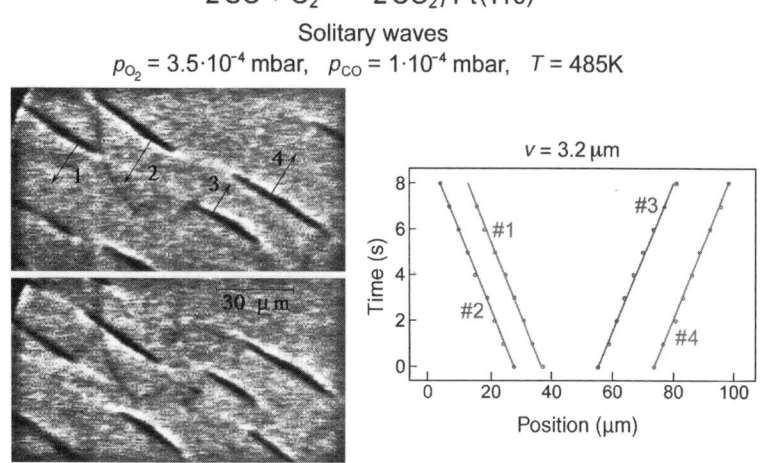

FIGURE 8.7. Solitary O waves propagating along the [0 0 1] direction on a Pt(1 1 0) surface during CO oxidation [21]. (See color insert.)

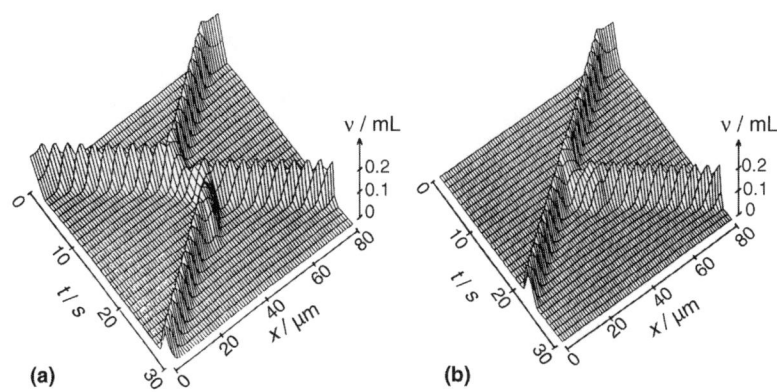

FIGURE 8.8. Theoretical (one-dimensional) profiles of solitary waves propagating along x-direction with time t. ν denotes the concentration of adsorbed O atoms causing the dark pulses in the PEEM images of Fig. 8.7 [22]. (a) Collision of two pulses propagating in opposite directions in a region of surface defects causes soliton-like behavior. (b) If a single pulse enters a defect zone, wave splitting may take place leading to two pulses propagating in opposite directions.

them reemerge from this collision, an effect denoted as soliton-like behavior. Moreover, occasionally one pulse is observed to emit another one propagating in the opposite direction (wave splitting). These effects have to be attributed to nonuniformities of the single-crystal surface. Under the assumption that the $1 \times 1 \rightarrow 1 \times 2$ structural transformation is not perfect everywhere (causing variations of the local oxygen sticking coefficient), these effects could all be reproduced by theory, as shown in Fig. 8.8 for the effects of soliton formation (a) and wave splitting (b) [22].

The aforementioned example is characteristic of a situation in which the steady-state (macroscopic) reaction rate is constant, but the reacting adsorbed species are nevertheless not uniformly distributed over the surface. Most common in this respect are spiral waves that are characteristic of excitable media, where the excitation not necessarily requires an external perturbation, but may often be identified with an inherent local variation of kinetic parameters such as a defect zone on the surface. The core of the spiral may then be formed by a local defect to which the spiral is

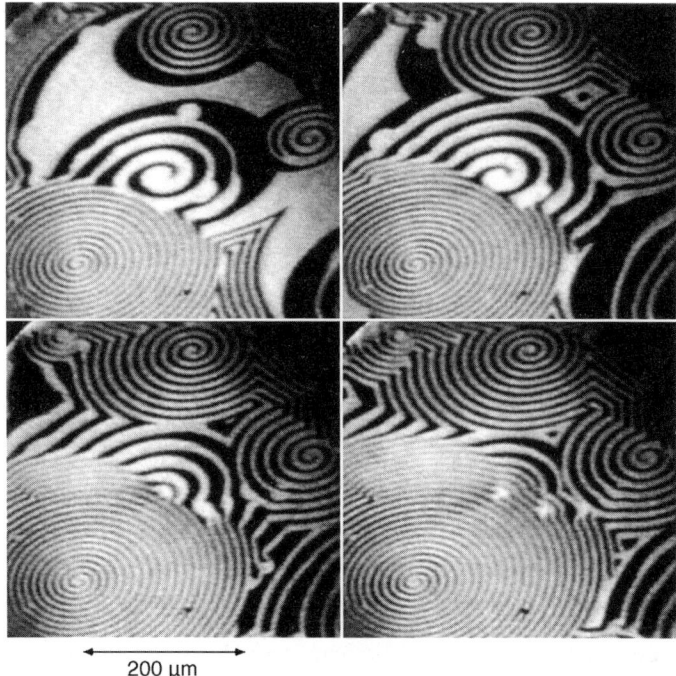

200 µm

FIGURE 8.9. A sequence of PEEM images from a Pt(1 1 0) surface during spiral wave formation in CO oxidation taken at intervals of 30 s. $T = 448$K, $p_{O_2} = 4 \times 10^{-4}$ mbar, $p_{CO} = 4.3 \times 10^{-5}$ mbar [24].

pinned and that determines its characteristic properties such as its wavelength and rotation period [23]. Mechanisms of the nucleation of surface defects were recently studied in detail [26]. Figure 8.9 shows a sequence of PEEM images from a Pt(1 1 0) surface during CO oxidation [24]. Not all spirals exhibit the same periods and wavelengths due to pinning of the cores to surface defects of varying size and kinetic properties. Extended defects (>10 µm) may even act as cores for multiarmed spirals. The waves propagate again with a speed of several µm/s, while on atomic scale about 10 CO_2 molecules per second and site are produced within the dark areas, demonstrating the quite different time scales on microscopic and mesoscopic length scales. Theoretical simulation of spiral wave formation was again performed by numerical

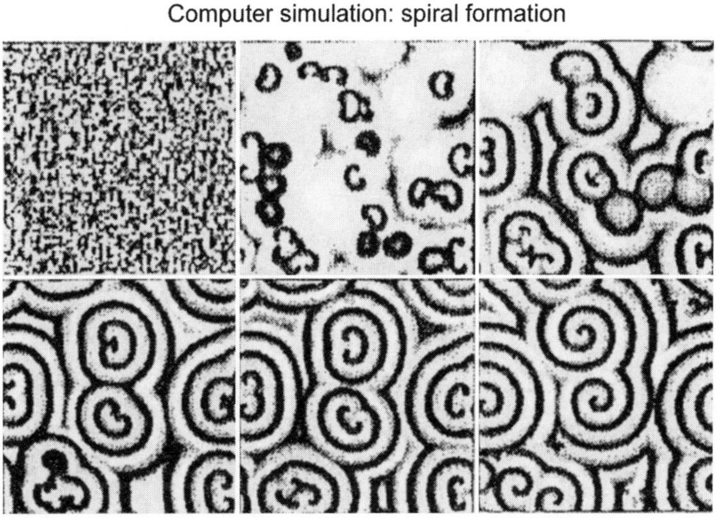

Computer simulation: spiral formation

FIGURE 8.10. Sequence of patterns leading to spiral wave formation starting with a disordered configuration of the adsorbates resulting from numerical solution of the PDEs modeling CO oxidation on Pt(1 0 0) [25].

solution of the quoted PDEs with properly chosen external parameters [25]. A sequence of resulting patterns is reproduced in Fig. 8.10: Starting from random distributions of the adsorbed species, nucleation of spiral cores developing to propagating waves takes place. Detailed theoretical modeling of the present system revealed that under certain conditions the cores are migrating around on the surface (meandering). Eventually, a different situation is found for slightly changed parameters. As shown in Fig. 8.11, developing spirals break up into a disordered state reminiscent of a turbulent fluid. That is why this state of spatiotemporal chaos is also denoted as chemical turbulence. Its experimental verification is reflected by the PEEM image reproduced in Fig. 8.12 [18].

8.4. MODIFICATION AND CONTROL OF SPATIOTEMPORAL PATTERNS

The propagating concentration patterns described in Section 8.3 can be affected in various ways.

Computer simulation: spiral turbulence

FIGURE 8.11. Slight variation of the control parameters leads to breakup of the spirals and emergence of chemical turbulence [25].

Alkali metals uniformly distributed over a surface often act as promoters in the kinetics of catalytic reactions. Their high mobility can give rise, on the other hand, to pattern formation. In the $H_2 + O_2$ reaction on a Rh(1 1 0) surface, for example, preadsorbed K atoms may condense reversibly into mesoscopic islands where they interact with O atoms and may then become subject of propagating reaction fronts [27,28]. Differences in the mobility and bonding strength of K on the O-rich and reduced surface regions are decisive factors for this type of concentration pattern formation.

Chemically inert surface atoms may also affect the kinetic properties of a surface beyond the atomic scale. In the case of CO

FIGURE 8.12. PEEM image showing the appearance of chemical turbulence in the CO oxidation on Pt(1 1 0) [18].

oxidation on Pt(1 1 0), the presence of a submonolayer of Au atoms reduces the sticking coefficient for oxygen as well as the diffusion coefficient for adsorbed CO if regarded on a mesoscopic length scale, which in turn affect the velocities of propagation of O and CO waves [29].

Figure 8.13 shows a PEEM image taken during catalytic CO oxidation at a Pt(1 1 0) surface on which about 5% of a monolayer of Au had been deposited on the left part, while the other part was pure Pt. Both the phenomenology and the dynamics (velocity of wave propagation) show pronounced differences on mesoscopic scale. That means the kinetic parameters averaged over the diffusion length again give rise to uniform behavior even if the atomic-scale structure is inhomogeneous. A straight wave front striking the boundary between the bare and Au-covered areas exhibits phenomena of refraction and reflection, as observed with light propagating in media with different optical refractive indices.

An extension of the aforementioned approach is achieved by fabricating surfaces with mesoscopic structures of a foreign

Au-covered Pt   Pt

FIGURE 8.13. PEEM image from a Pt(1 1 0) surface on which the left part was covered by 5% by a monolayer of Au atoms [34].

(inert) material. Photolithographic techniques as developed for microelectronics offer a convenient possibility to prepare such samples. It is known that the formation and propagation of nonlinear waves may be markedly affected by the spatial boundary conditions in that, for example, certain modes are selected or the motion through narrow channels is suppressed [30,31]. This principle has been verified with a Pt(1 1 0) surface onto which a mask of titanium had been deposited [32]. This material oxidizes and is then completely inactive in the CO oxidation reaction, which is therefore restricted to the bare Pt regions. A PEEM image from such a sample during CO oxidation exhibiting the propagation of concentration waves is reproduced in Fig. 8.14. While the letters F and I are in uniform states, spirals and propagating pulses develop on the letter H. A rich variety of

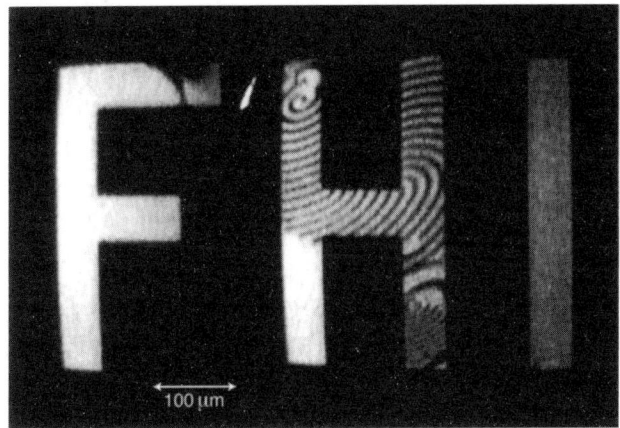

FIGURE 8.14. Pattern formation on a microstructured Pt(1 1 0) surface during CO oxidation [32].

different phenomena has been observed in the meantime with similar systems [33–35].

Transient modification of a local area on an adsorbate-covered surface may be achieved by focusing a laser beam onto a small spot of about 80 μm on the surface, and by adjusting this spot by computer-controlled galvanometer mirrors within 5 μm and on a timescale of 1 ms. In this way, the local temperature can typically be increased by about 3K [36]. In the example shown in Fig. 8.15,

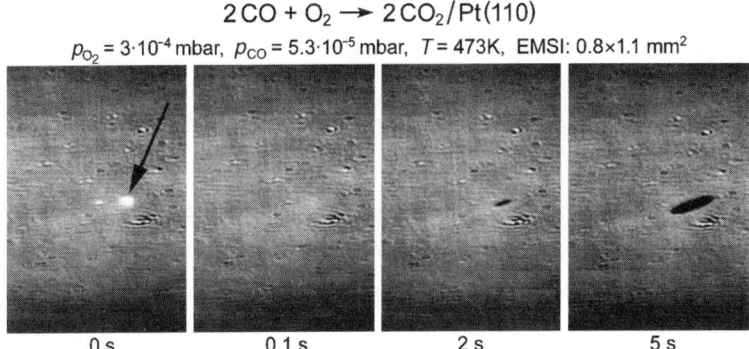

FIGURE 8.15. Triggering of a concentration wave in CO oxidation on Pt(1 1 0) at the bright spot irradiated by a short laser pulse [37].

a Pt(1 1 0) surface during steady-state CO oxidation was irradiated at $t=0$ with a laser shot that causes local CO desorption, so that oxygen adsorption takes place where a reaction is ignited that propagates into the CO-covered surrounding [37]. The elliptical shape of this growing pattern has again to be attributed to the anisotropy of surface diffusion on the Pt(1 1 0) surface.

Instead of irradiating just a single spot, the laser beam may also be moved across the surface, thus forming transient patterns with altered adsorbate composition. Thus, the overall reaction rate may also be affected [38]. Apart from the initiation of reaction waves [39], already moving waves may be "dragged" and thus affected in their movement [40].

The kinetics of periodically forced oscillations in the CO oxidation on Pt(1 1 0) leading to the effects of entrainment and quasi-periodicity have been discussed in Section 7.3. The associated changes of the concentration patterns were investigated more recently [41]. A rich variety of complex patterns is observed. For example, near harmonic resonance with the forcing, intermittent turbulence characterized by localized turbulent "bubbles" on a homogeneously oscillating background appears, while for the 2:1 subharmonic range, irregular oscillatory stripes and cluster patterns are found.

Apart from external periodic forcing, complex patterns may also be modified by global delayed feedback. The principle of this approach for our systems is illustrated in Fig. 8.16 [42]: The brightness of a whole PEEM image (reflecting the overall reaction rate) is determined by integration and serves as input for a feedback loop in which one of the control parameters (in our case the CO pressure) is altered with a predescribed strength and delay time. (For example, if the image is too dark with respect to a chosen standard value, the CO valve is opened slightly more to reduce the O coverage on the surface). If one starts from a state of spatiotemporal chaos or chemical turbulence as reproduced in the center of Fig. 8.17, depending on the choice of the control parameters,

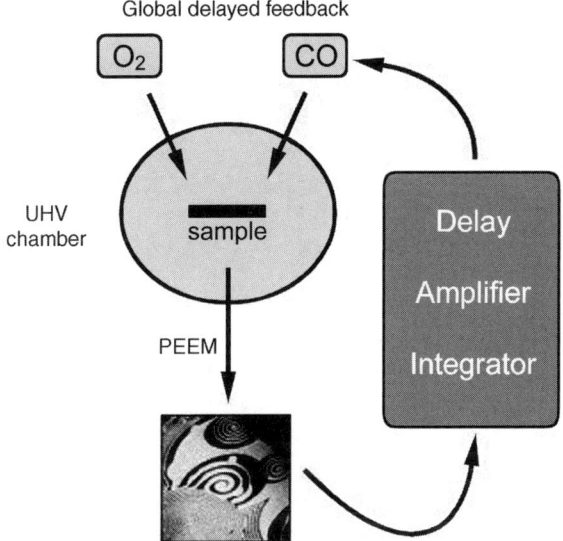

FIGURE 8.16. Principle of controlling pattern formation by global delayed feedback [42].

namely, strength and delay time of the feedback, quite different new patterns may be created, which are imaged in Fig. 8.17 and are also reproduced by theory. These range from various types of turbulence to larger "clusters" and cellular structures to uniformly oscillating background.

The next step of sophistication consists in the introduction of a novel nonuniform coupling scheme where the feedback loop is sensitive to the length scales of patterns emerging in the medium [43]. This approach can be regarded as a first step toward flexible feedback protocols for guided self-organization in nonequilibrium systems, and one might speculate that nature operates along similar strategies to create optimal structures.

8.5. THERMOKINETIC EFFECTS

The effects caused by relatively small local perturbations of the surface temperature, as discussed in Section 8.4, suggest that

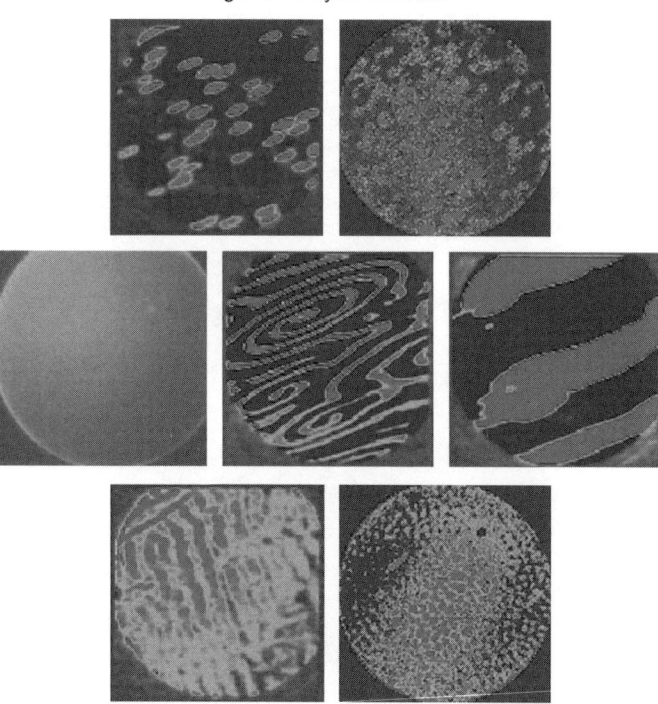

FIGURE 8.17. Summary of controlled formation of patterns in CO oxidation on Pt(1 1 0) by global delayed feedback with varying delay times and strengths of feedback [42]. (See color insert.)

temperature changes caused by the finite reaction enthalpy of a catalytic reaction may manifest themselves as well in the spatiotemporal self-organization. Such effects are likely to occur with systems operated at elevated (i.e., up to atmospheric) pressures where substantial temperature changes may accompany the reaction.

Local temperature variations can conveniently be recorded by IR thermography [46], and there are several reports in the literature on the investigation of such effects with the CO oxidation on supported catalysts [47–49]. These experiments demonstrated the importance of thermal coupling between different catalyst pellets and its influence on the degree of synchronization. By using a thin evaporated Pt film, the propagation of reaction waves could be demonstrated [50]. In another study, a Pt wire was used that was solely heated by the reaction enthalpy released. Here the oscillatory behavior changed substantially if the wire loop was cut into two parts, again demonstrating the relevance of the transport in such systems [51].

The coupling of chemical rate oscillations to temperature variations and then to thermal expansion causing deformation of a thin film was demonstrated with a 200 nm thick Pt single crystal exposing the (1 1 0) surface [52]. As shown in Fig. 8.18a, the thin sample was placed over a hole in a bulk Pt plate. Kinetic oscillations with a period of about 5 s developed at a O_2 partial pressure of 5×10^{-3} mbar and caused temperature variations of the central part by about 30 K. These in turn caused periodic deformations of the thin foil, as reproduced in Fig. 8.18b, as measured by integrating the reflected light intensity from the area marked in the third frame of Fig. 8.18c. Part (c) of this figure shows images from the deformations of the catalyst's surface at the four instants marked in Fig. 8.18b. Detailed theoretical analysis of the interplay between chemical reaction, heat evolution, and

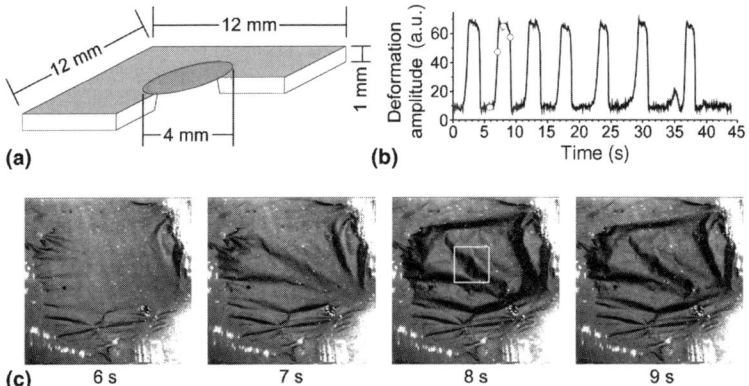

FIGURE 8.18. Thermomechanical instability of a thin (1 1 0) Pt catalyst during oscillatory CO oxidation [52]. (a) Catalyst and support geometry. (b) Deformation amplitude as a function of time. (c) Images ($4.4 \times 4.4 \, \text{mm}^2$) from the catalyst's surface at various times marked in (b).

mechanical deformation initiated by thermal expansion provided profound rationalization of the mechanism underlying these "heartbeats" of a catalyst.

Systematic studies on thermokinetic wave propagation have been reported for various reactions, and different approaches have been reported to implement nonisothermal effects in the theoretical modeling [53]. Even if the details of the reaction mechanism are less well understood, the basic features of spatiotemporal pattern formation with such systems can often be modeled successfully, because the decisive effects can be approximated by a heat balance equation in which the chemistry is reduced to single variable and surface diffusion of adsorbates is neglected [54,55].

8.6. Pattern Formation on Microscopic Scale

The small particles of "real" supported catalysis as well as all objects forming the subject of nanotechnology have dimensions that are much smaller than the typical patterns found with isothermal reaction–diffusion systems on extended uniform surfaces. Typical nonlinear effects with supported catalysts are usually of thermokinetic origin for which only the properties averaged over macroscopic length scales are of significance, as discussed in Section 8.5. The situation changes, however, if processes occurring near atomic scale become the subject of closer inspection.

Such effects can, for example, be observed by applying the field ion microscope (FIM) to sharp tips exposing different crystal planes. Quite interesting observations in this connection were made with the $H_2 + O_2$ reaction on Pt [56]. With an extended Pt(1 0 0) surface under the applied conditions, no kinetic oscillations but only bistable behavior are found. If the same plane is, however, exposed as a small area on a field emitter tip, then under certain conditions oscillatory behavior is observed. This

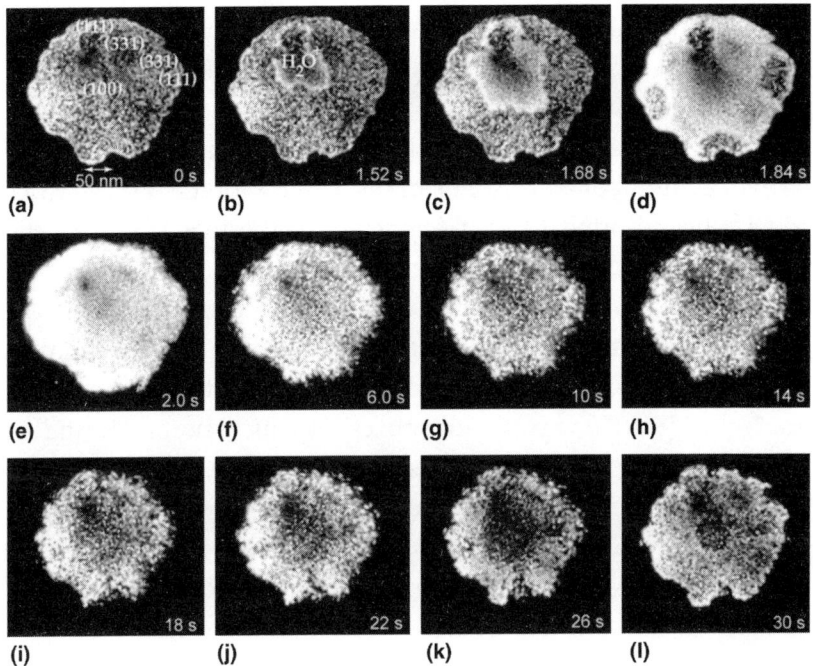

FIGURE 8.19. Series of field ion microscopy images from a Pt tip during the $H_2 + O_2$ reaction at $T = 300\,K$, $p_{H_2} = 6 \times 10^{-2}$ mbar, and $p_{O_2} = 5 \times 10^{-2}$ mbar. Imaging gases are O_2 and the H_2O formed by the reaction [56].

phenomenon is illustrated by the sequence of FIM images reproduced in Fig. 8.19. The central Pt(1 0 0) region becomes periodically activated by reaction–diffusion waves triggered from adjacent high-index crystal planes. It is important to distinguish this effect from the well-known phenomenon of "spillover" that is simply due to diffusion of reacting species from one part of the surface to the other. Here, it is the coupling of reaction and diffusion that causes the propagation of nonlinear concentration waves under conditions far from equilibrium and thus affects the overall reactivity. Such effects are presumably of general importance for the reactivity of small supported catalyst particles whose dimensions are comparable to those of the FIM tips. Their activity is usually considered to be just a superposition of the contributions of the various crystal planes. Computer simulations for models representing such small particles have demonstrated the limitations of this linear approach and emphasize the importance of the sketched nonlinear effects [57]. On the experimental side, it was demonstrated that coverage fluctuations on small catalyst particles alter their macroscopic catalytic behavior, for example, macroscopically observable bistabilities disappear with decreasing particle size [58].

Even with extended uniform surfaces, the operation of attractive interactions between the reacting species may lead to the formation of structures on the nanoscale with very sharp interfaces and even uphill diffusion against an existing concentration gradient [59]. Generally, structures of this type are found in reactive monolayers acting as soft matter systems where energetic interactions between molecules come into play and where chemical reactions and diffusion may interfere with phase transitions [60]. It is felt that such nonequilibrium soft matter microstructures are of fundamental significance for biological systems.

References

1. I. R. Epstein and J. A. Pojman, *An Introduction to Nonlinear Chemical Dynamics*, Oxford University Press, 1998.
2. R. Kapral and K. Showalter (eds.), *Chemical Waves and Pattern Formation*, Kluwer, 1995.
3. M. Ehsasi, O. Frank, J. H. Block, and K. Christmann, *Chem. Phys. Lett.* **165** (1990) 115.
4. M. Eiswirth and G. Ertl, *Phys. Rev. Lett.* **60** (1988) 1526.
5. K. Krischer, *J. Electroanal. Chem.* **501** (2001) 1.
6. A. Turing, *Phil. Trans. R. Soc.* **237** (1952) 37.
7. V. Castets, E. Dulos, J. Boissonade, and P. de Kepper, *Phys. Rev. Lett.* **64** (1990) 2953.
8. Q. Ouyang and H. L. Swinney, *Nature* **352** (1991) 610.
9. (a) S. Sick, S. Reinker, J. Timmer, and T. Schlake, *Science* **314** (2006) 1447; (b) P. K. Maini, R. E. Baker, and C. M. Chuong, *Science* **314** (2006) 1397.
10. N. Mazouz and K. Krischer, *J. Phys. Chem. B* **104** (2000) 6081.

11. G. Flätgen, K. Krischer, B. Pettinger, K. Doblhofer, H. Junkes, and G. Ertl, *Science* **269** (1995) 668.
12. Y. L. Li, J. Oslonovitch, N. Mazouz, F. Plenge, K. Krischer, and G. Ertl, *Science* **291** (2001) 2395.
13. D. Bizzotto and A. Nelson, *Langmuir* **14** (1998) 6269.
14. W. Engel, M. Kordesch, H. H. Rotermund, S. Kubala, and A. von Oertzen, *Ultramicroscopy* **36** (1991) 148.
15. H. H. Rotermund, G. Haas, R. U. Franz, R. M. Tromp, and G. Ertl, *Science* **270** (1995) 606.
16. G. Ertl, *Science* **254** (1991) 1750.
17. R. Imbihl and G. Ertl, *Chem. Rev.* **95** (1995) 697.
18. S. Jakubith, H. H. Rotermund, W. Engel, A. von Oertzen, and G. Ertl, *Phys. Rev. Lett.* **65** (1990) 3013.
19. H. Levine and X. Zhou, *Phys. Rev. E* **48** (1993) 50.
20. A. von Oertzen, H. H. Rotermund, A. S. Mikhailov, and G. Ertl, *J. Phys. Chem. B* **104** (2000) 3155.
21. H. H. Rotermund, S. Jakubith, A. von Oertzen, and G. Ertl, *Phys. Rev. Lett.* **66** (1991) 3083.
22. M. Bär, M. Eiswirth, H. H. Rotermund, and G. Ertl, *Phys. Rev. Lett.* **69** (1992) 945.
23. J. P. Keener and J. J. Tyson, *Physica D* **21** (1986) 300.
24. S. Nettesheim, A. von Oertzen, H. H. Rotermund, and G. Ertl, *J. Chem. Phys.* **98** (1993) 9977.
25. M. Bär, N. Gottschalk, M. Eiswirth, and G. Ertl, *J. Chem. Phys.* **100** (1994) 1202.
26. H. Wei, G. Lilienkamp, and R. Imbihl, *J. Chem. Phys.* **127** (2007) 024703.
27. H. Marbach, M. Hinz, S. Günther, L. Gregoratti, A. Kiskinova, and R. Imbihl, *Chem. Phys. Lett.* **364** (2002) 364.
28. H. Marbach, S. Günther, T. Neubrand, R. Hoyer, L. Gregoratti, M. Kiskinova, and R. Imbihl, *J. Phys. Chem. B* **108** (2004) 15182.
29. K. Asakura, J. Lauterbach, H. H. Rotermund, and G. Ertl, *Phys. Rev. B* **50** (1994) 8043.
30. A. Balloyantz and J. A. Sepulchre, *Physica D* **49** (1991) 52.
31. J. Enderlein, *Phys. Lett.* **156** (1991) 429.
32. M. D. Graham, I. G. Kevrekidis, K. Asakura, J. Lauterbach, K. Krischer, H. H. Rotermund, and G. Ertl, *Science* **264** (1994) 80.
33. S. Y. Shvartsman, E. Schütz, R. Imbihl, and I. G. Kevrekidis, *Catal. Today* **70** (2001) 301.

34. J. Lauterbach, H. H. Rotermund, P. B. Rasmussen, M. Bär, M. D. Graham, I. G. Kevrekidis, and G. Ertl, *Physica D* **123** (1998) 493.
35. M. Lawin, V. Johannek, A. W. Grant, B. Kasemo, J. Libuda, and H. J. Freund, *J. Chem. Phys.* **122** (2005) 084713.
36. J. Wolff, A. G. Papathanasiou, I. G. Kevrekidis, H. H. Rotermund, and G. Ertl, *Science* **294** (2001) 134.
37. J. Wolff, A. G. Papathanasiou, H. H. Rotermund, G. Ertl, X. Li, and I. G. Kevrekidis, *J. Catal.* **216** (2003) 246.
38. A. G. Papathanasiou, J. Wolff, I. G. Kevrekidis, H. H. Rotermund, and G. Ertl, *Chem. Phys. Lett.* **358** (2002) 407.
39. J. Wolff, A. G. Papathanasiou, H. H. Rotermund, G. Ertl, M. A. Katsoulakis, X. Li, and I. G. Kevrekidis, *Phys. Rev. Lett.* **90** (2003) 148301.
40. J. Wolff, A. G. Papathanasiou, H. H. Rotermund, G. Ertl, X. Li, and I. G. Kevrekidis, *Phys. Rev. Lett.* **90** (2003) 018302.
41. M. Bertram, C. Beta, H. H. Rotermund, and G. Ertl, *J. Phys. Chem. B* **107** (2003) 9610.
42. M. Kim, M. Bertram, M. Pollmann, A. von Oertzen, A. S. Mikhailov, H. H. Rotermund, and G. Ertl, *Science* **292** (2001) 1357.
43. C. Beta, H. G. Moula, A. S. Mikhailov, H. H. Rotermund, and G. Ertl, *Phys. Rev. Lett.* **93** (2004) 188302.
44. J. Christoph, P. Strasser, M. Eiswirth, and G. Ertl, *Science* **284** (1999) 291.
45. J. Lee, J. Christoph, T. Noh, M. Eiswirth, and G. Ertl, *J. Chem. Phys.* **126** (2007) 144702.
46. V. Barelko, I. I. Kurachka, A. G. Merzhanao, and K. G. Shkadinskii, *Chem. Eng. Sci.* **33** (1978) 805.
47. J. C. Kellow and E. E. Wolf, *Chem. Eng. Sci.* **45** (1990) 2597.
48. C. C. Chan, E. E. Wolf, and H. C. Chang, *J. Phys. Chem.* **97** (1993) 1055.
49. H. U. Onken and E. Wicke, *Ber. Bunsenges.* **90** (1986) 976.
50. J. P. Dath and J. P. Dauchot, *J. Catal.* **115** (1989) 593.
51. D. K. Tsai, M. B. Maple, and R. K. Herz, *J. Catal.* **113** (1988) 453.
52. F. Cirak, J. E. Cisternas, A. M. Cuitino, G. Ertl, P. Holmes, I. G. Kevrekidis, M. Ortiz, H. H. Rotermund, H. Schunack, and J. Wolff, *Science* **300** (2003) 1932.
53. G. Ertl, in: *Handbook of Heterogeneous Catalysis* (eds. G. Ertl, H. Knözinger, J. Weitkamp, and F. Schüth), Wiley, 2008, p. 1492.
54. Yu. E. Volodin, V. N. Zuyagin, A. N. Ivanova, and V. Barelko, *Adv. Chem. Phys.* **77** (1990) 551.
55. U. Middya, D. Luss, and M. Sheintuch, *J. Chem. Phys.* **100** (1994) 3568.

References

56. V. Gorodetskii, J. Lauterbach, H. H. Rotermund, J. H. Block, and G. Ertl, *Nature* **370** (1994) 277.
57. (a) V. P. Zhdanov and B. Kasemo, *Surf. Sci.* **496** (2002) 251; (b) V. P. Zhdanov and B. Kasemo, *J. Catal.* **214** (2003) 121.
58. V. Johanek, M. Laurin, A. W. Grant, B. Kasemo, C. R. Henry, and J. Libuda, *Science* **304** (2004) 1639.
59. A. S. Mikhailov and G. Ertl, *Chem. Phys. Lett.* **238** (1995) 104.
60. A. S. Mikhailov and G. Ertl, *Chem. Phys. Chem.* **10** (2009), 86.

Index

activated adsorption   13
active sites   21, 105, 135
adsorption energy   5
alkali promoter   135
angular distribution of desorbing particles (ESDIAD)   89
anisotropic surfaces   14

Belousov-Zhabotinsky reaction   160
bifurcation theory   163
Born-Oppenheimer approximation   79
bulk phases   43

C7 sites   125
car exhaust   139
carbonyl compounds   9
catalyst   103
catalytic activity   107
chemical turbulence   189
chemical vapor deposition (CVD)   45
chemicurrents   86, 87
chemiluminescence   80
chemisorption energy   4
chemisorption   4, 9
chemistry with a hammer   59
Clausius-Clapeyron equation   5
closed system   175
CO chemisorption   9
coadsorption   111
collision-induced surface reactions   59
competitive adsorption   112
consecutive reactions   117

continuous flow   103
cooling in desorption   65
cooperative adsorption   111
coordination chemistry   9
coverage   11

density functional theory (DFT)   10
desorption induced by electronic transitions (DIET)   88
desorption   4
deterministic chaos   168
DFT calculations   137
diffusion coefficient   14, 16
diffusion length   152
DIMET   72, 73, 94
direct elastic scattering   52
direct inelastic scattering   54
direct valence excitation   94
displacive reconstruction   35
dissociative adsorption   7, 13, 21, 54
dynamic phase diagram   171
dynamics   51

early barrier   56
electrochemical cells   29
electrochemical pulse techniques   28
electrochemical reaction   2
electrochemical surface machining   31
electrochemical systems   179
electrochemical Turing patterns   180
electronic excitations   79, 80, 86
electronic promoter   125

Reactions at Solid Surfaces. By Gerhard Ertl
Copyright © 2009 John Wiley & Sons, Inc.

electron-stimulated desorption 88
Eley-Rideal (ER) mechanism 109, 110
energetics of chemisorption 4
energy exchange 52, 69
epitaxy 44
exoelectron emission 81 ff.
extrinsic precursor 11

faceting 41
field ion microscope (FIM) 14, 198
forced oscillations 169
Frank-van der Merwe growth 46

gas-phase coupling 177
global coupling 185
global delayed feedback 194

Haber-Bosch process 124
harpooning 83
heartbeats 198
heat conductance 177
Hedvall effect 42
heterogeneous catalysis 1, 103
hot adatoms 60, 62, 63
hot electrons 72, 75, 79, 94

internal electron excitation 86
intrinsic precursor 13
isothermal wave patterns 183

kinetic Monte Carlo approach 116
kinetic oscillations 38
kinetics of catalytic reactions 113
kinetics of chemisorption 11

Langmuir approach 114
Langmuir kinetics 133
Langmuir-Hinshelwood (LH)
   mechanism 109, 111, 139, 140, 147
late barrier 56
Lennard-Jones diagram 8, 54
Lotka-Volterra model 162

materials gap 4, 145
mean field approximation 145
mesoscopic scale 175
mesoscopic structures 191
metastable structures 26
MGR model 88
microkinetic modeling 133, 135
microkinetics 116

microstructures 30
missing row structure 37, 38
Mittasch catalyst 114
molecular beam epitaxy (MBE) 45
molecular beam experiments 58, 64, 140
monoatomic step 22

nanotechnology 1, 2
nonadiabatic coupling 79
nonadiabatic surface reaction 81
nonlinear dynamics 159, 160
nucleation 25

open systems far from equilibrium 175
ordinary differential equations
   (ODEs) 160
oscillatory kinetics 159, 160
Ostwald process 117
Ostwald ripening 26
oxidation of carbon monoxide 139
oxidation of hydrogen 149
oxide-nucleation 44

parallel reactions 117
partial differential equations (PDEs) 152, 161
periodically forced oscillations 194
phase diagram 24
phase portraits 168
phase transition 36
photochemical reactions 94 ff.
photoemission electron microscopy
   (PEEM) 183
poison 107
potassium promoter 131
potential energy diagram for ammonia
   synthesis 132
precursor model 11, 60
pressure gap 4, 105, 125, 145
promoter 113

reaction coordinate 51
reaction-diffusion (RD) model 154
reactive scattering 54
remote triggering 182
rotational rainbow 68

scanning tunneling microscopy
   (STM) 141 ff.
scattering at surfaces 52
segregation 42

selectivity   117
self-assembly   26
self-organization   26, 175, 176
single-crystal surfaces   3
solitary waves   185
spatiotemporal chaos   189
spillover   199
spinodal decomposition   27, 28
spiral waves   187
state-resolved molecular beam
   experiments   66
steady-state kinetics   147
sticking coefficient   11, 130
strain-induced long-range order   40
strange attractor   168
Stranski-Krastanov growth   46
structural promoters   125
structural transformation   37
structure insensitive   24
structure sensitive   24
subsurface atoms   42
sum frequency spectroscopy (SFG)   70
surface diffusion   9, 16, 103, 176
surface femtochemistry   79
surface nitrides   127
surface photochemistry   94
surface reaction   103

surface reconstruction   33, 38
synthesis of ammonia on iron   123
synthesis of ammonia on ruthenium   134

target patterns   185
temperature-programmed reaction
   spectroscopy (TPRS)   109
thermal accommodation   64
thermal desorption spectroscopy (TDS)   6, 137
thermal desorption   13, 64
thermionic energy conversion   81
thermokinetic effects   195
tip-induced desorption   92
transition state theory (TST)   51
transition state   137
transition to chaos   168
trapping desorption   54
Turing patterns   178
turnover frequency (TOF)   109
two-dimensional phases   24
two-photon photoelectron
   spectroscopy   93

Volmer-Weber growth   46

wave propagation   151